Molecular Origins of Brain and Body Geometry

Antonio Lima-de-Faria

Molecular Origins of Brain and Body Geometry

Plato's Concept of Reality is Reversed

Springer

Prof. Dr. Antonio Lima-de-Faria
Lund University Department of Cell and Organism Biology
Biology Building
Sölvegatan 35
SE- 223 62
Lund
Sweden

Private address:
Trumslagaregränd 10
SE-226 39 Lund
Sweden

ISBN 978-3-319-06055-2 ISBN 978-3-319-06056-9 (eBook)
DOI 10.1007/978-3-319-06056-9
Springer Cham Heidelberg New York Dordrecht London

Library of Congress Control Number: 2014950216

Printed on acid-free paper

Springer is part of Springer Science+Business Media (www.springer.com)

Acknowledgments

Med. Kand. Johan Essen-Möller is to be thanked for an unfailing contribution in converting the manuscript into computer work which he carried out with the utmost accuracy.

Special thanks are due to the staff of the Biology Library, Lund University, for the retrieval of scientific works: Librarian Ph.D. Kristina Arnebrant, Librarian Ph.D. Bodil Enoksson, Librarian B.Sc. Johnny Jönsson and Ms. Sabina Swahn.

Thanks are also due to the Editors of Springer who accepted the two manuscripts without reservation on the sole basis of an outline and a sample chapter: Senior Publishing Editor Dr. Meran Owen ("Molecular Geometry of Body Pattern in Birds", Springer 2012), Editorial Director Dr. Hanna Hensler-Fritton, Editor Dr. Verena Penning and Dr. Lars Koerner (the present work).

I would like to thank my colleagues for valuable support: Professor Dr. Harry Scherthan, Institute of Radiobiology, University of Ulm, Munich, Germany; Doctor Josep del Hoyo, Lynx Edicions, Barcelona, Spain and Professor Dr. Walter Sachsse, Institut für Molekulargenetik, University of Mainz, Germany.

"Permission Forms" were sent by Air Mail to all Publishers to request permission to use the figures included in this work. Most publishers did not demand a fee, some kindly made a 60% reduction. Other publishers demanded a fee which was payed immediately as the book went into press. The "Sources of Figures" are at the end of every chapter.

Contents

1 Introduction

1.1 Unintentionally the Research on the Molecular Geometry of Vertebrates Led to Plato

It was never the intention, when starting this work, to come to a conclusion that would have philosophical implications.

Plato's main contention (c. 429–347 BC), was that the abstract "forms" conceived by the mind were changeless and represented the only reality. In opposition the outer world consisted of objects that were unreal, imperfect and fleeting (Blackburn 1996). This was a distinction of great importance but from the start it was contested.

How does one dare to reverse Plato's main contention being a molecular biologist? The reason is simple.

At every stage of its development, philosophy was influenced by the science of its day. It collected experimental information that led it to produce novel concepts. Aristotle (384–322 BC) wrote his treatise on "Ethics" based on the newly established natural sciences. Later, the philosopohy of René Descartes (1596–1650) was influenced by the dissection of the human body carried out for the first time in anatomical theatres, spread over Europe from Italy to Sweden. It led him to locate the "soul" in the pineal gland. In our time Popper (1902–1994), from the Vienna school of philosophy, criticized Darwinism considering it metaphysical (Popper 1976).

Whenever new evidence emerges in biology, it has implications for philosophy, whether we like it or not.

An example at hand is the present expansion of biotechnology. The fusion of human with plant cells, considered impossible at one time, was obtained by Dudits et al. (1976). It led to human chromosomes dividing in a plant-human cytoplasm, an event controlled by labelling the chromosomes with radioisotopes and other techniques. These experiments were followed by the transfer of human genes into mice (Kuehn et al. 1987) and later by introducing human genes into plants to obtain better medicines. The cloning of the sheep "Dolly" (Wilmut et al. 1997) led several laboratories to clone human embryos for therapeutic purposes (Tachibana et al. 2013).

The ethical implications of these experiments are paramount. Results in molecular biology cannot easily be separated from central philosophical questions.

1.2 A Plethora of Genes Allows to Map Body Pattern and Brain Functions in Vertebrates

This work deals solely with the biology of vertebrates, from fishes to humans. Their structures and functions are largely decided by their many genes. These have been: (1) identified, (2) isolated in the test tube, (3) localized in specific chromosomes, (4) DNA sequenced and (5) had their expression (consisting of molecular cascades) well documented. The proteins that these genes produce have, in some cases, been amino acid sequenced. Hence we are dealing with an accurate mapping of body patterns and functions that previously escaped us.

The information on gene action not only encompasses the outer body configuration but that of the brain. Besides it is not a question of single genes but of whole gene batteries. Not less than 2,500 genes are necessary for eye formation and more than a thousand opsin proteins have been identified that are light sensors in vertebrates.

1.3 The Impressive Colored Markings of Birds Combined with Genetic Evidence Were the Starting Point that Led to Molecular Geometry

Among the vertebrates, birds have drawn special attention from naturalists due to their rainbow colors and exquisite body configurations.

The patterns built by the feathers of the bird's body have been seen by ornithologists as "amazing curiosities" or

A. Lima-de-Faria, *Molecular Origins of Brain and Body Geometry*,
DOI 10.1007/978-3-319-06056-9_1, © Springer International Publishing Switzerland 2014

"bizarre events" demanding little attention from the genetic point of view.

However, when the over 8,000 species were analysed in detail and the genes responsible for the color pigmentation, and feather formation, were investigated the following main features emerged: (1) Concentric circles of different colors appeared around the eye and extended to the head, breast, wings and tail, covering the whole body with a circular pattern. (2) Parallel stripes ran over the bird's body following the head-tail axis, whereas others had a 90° distribution to this orientation. (3) Protruding crests were found to be semicircular and were accompanied by concentric semicircles of different bright colors. (4) The tails were equally impressive showing arrays of yellow and blue concentric rings. (5) The bill and legs were not left outside this geometry—they exhibited stripes of different colors which were parallel to each other. (6) These geometric patterns occurred repeatedly appearing with equal clarity in the most unrelated avian species. This is not surprising since a given DNA sequence can function as a repressor or an activator depending on the DNA's sole molecular properties as is the case with the *svb* gene (Kondo et al. 2010).

From the beginning nothing seemed to make sense. But there are no "curiosities" in nature. One of the basic tenets of genetics is that every natural trait is the product of gene action, interacting with the environment, and taking place at a well-defined stage of the development of the organism.

None of these surprising bird features were found to be fortuitous. They appeared to be an integral part of a rigid order directed by simple genes and molecular cascades occurring at various cellular levels. The figures appearing on the feathers built a novel geometry formed by pigments whose chemical composition is known. It could be ascertained that proteins and other molecules guided the pigments to well-defined locations on the feathers revealing that this geometry had a molecular basis (Lima-de-Faria 2012).

1.4 The Molecular Geometry Was Searched Across the Other Vertebrates

When a phenomenon becomes evident in a particular group of animals, a question arises. How general is it?

Birds, with their pageantry of colors bewilder the observer. Was their molecular geometry a case apart or a general phenomenon common to all vertebrates? Could its genetic basis also be extended to other animal orders?

It seemed that this effort was deemed to failure. Most vertebrates that we are familiar with do not show any obvious traits that can be compared to the painting palette of birds and their figures of geometry. Elephants are well known to have an even brown color, humans have no stripes, seals are mostly grey, frogs tend to be green and herrings are silvery-grey. But first appearances delude.

A careful analysis of the body pattern of vertebrates from fishes to apes, revealed brighter colors and geometric patterns even more distinct and revealing than those of the avian body.

In birds, the information from genes and molecular cascades was well established, but limited, because the DNA of birds has not been extensively studied. On the other hand, in other vertebrates, especially in humans and other mammals such as mice, the genetic data are more extensive and profound allowing to establish the molecular geometry on a more secure ground.

Tropical fishes are as startling in their colors and geometric circles as peacocks. Tortoises are covered with the most regular triangles, squares and concentric circles that can be green, brown or yellow. Scarlet bands are placed side by side of black ones along the body of snakes. Zebras and giraffes have patterns which are lessons in geometry, with their transversal and longitudinal stripes, their circles and other geometric figures. Monkeys, like the mandrills, have a spectacularly colored face—scarlet nose with blue parallel flanges and yellow beard.

All this geometry turns out to be also highly molecular. The genes have been DNA sequenced. Besides they cover most organs. They not only deal with the coloration of the body, but with the development of the brain, the formation of the eye, and the embryonic process. Most significant is that their action is interrelated since the same basic molecules decide the fate of quite different organs. An even more precise scenario of molecular events unravels in the vertebrates.

1.5 What Is a Vertebrate?

The vertebrates (backboned animals) started their evolution when the fish appeared about 440 million years ago (Barnes 1980; Burnie 2004). The formation of a bony skeleton in the interior of the spongy body of the invertebrates was a significant innovation. It gave the whole structure a stabilizing frame.

The main features of this group are as follows: Fish are mainly aquatic possessing gills, paired and unpaired fins and are usually covered with scales. Most fishes have a well-developed vertebral column, but hagfish lack well-defined vertebrae. A significant physiological trait is that they cannot control their body temperature. There are about 24,500 species.

Next in evolution appeared the amphibians. There are three living groups: salamanders, frogs and caecilians. What really unites the three orders is that they are intermediate between the aquatic fishes and the terrestrial reptiles comprising about 5,000 species. Like the fishes they are unable to regulate their body temperature and prefer aquatic environments. The eggs of amphibians lack both a shell and the embryonic membranes.

The reptiles arrived later. The sharp difference between amphibians and reptiles is that the latter exhibit a series of adaptations to life on land. They can inhabit moisty and dry environments and have a well developed excretory system (7,984 species).

The birds are easy to characterize. They have feathers, wings, only two legs and a bill, are warm-blooded and lay eggs, but have still much in common with reptiles. Birds are characterized by an increased cerebral and sensory development being the most common of the terrestrial vertebrates, outnumbering the number of mammal species twofold (about 9,000 vs. 4,475). But evolution has, in other respects, not been kind to the birds. They are toothless, their wings have no prehensile fingers and the penis is rare (copulating by means of an ineffective cloaca).

1.6 No Trait in Vertebrates Is Random

The last vertebrate group to emerge in evolution were the mammals. Hair is the most obvious and unique structure of this group. Mammals, like birds, have their heat generated from inside the body by continuous metabolic processes. Most mammals have four legs (each with five toes) a feature common to most terrestrial vertebrates. However, the arrangement of their limbs and modes of locomotion differ extensively from the other groups. The limbs of mammals are held directly beneath the body and move in a plane parallel to the long axis of the body. Also unique is the face, with its fleshy cheeks and lips, accompanied by muscles surrounding the mouth and eyes. Last, but not least, the presence of mammary glands is an obligatory feature of females.

Macdonald (2002), when summarizing the main features of mammals and pointing out their most diverse types of structure, ranging from shrews to whales, ends with a significant remark: “No facet of these varied lives is random; they are diverse but not in disarray”. This coherence of organization is not only true of the mammals but also of the other vertebrates.

What follows is an enquiry into the geometric figures displayed by the diverse classes of vertebrates to find out what they teach us concerning their perilous evolution. Moreover, it turns out that their brains have also their own geometry.

1.7 The Unifying Power of Geometry

Many questions remain unanswered and others are not yet even formulated. We have reached an impasse in the study of the brain. The following pages only describe how the geometry of matter appears to be related to that of the body and to that of the brain and how these in turn interact with each other in novel ways. No immediate solutions are at hand.

The reason why geometry has led the way in this inquiry is due to the fact that where there is geometry there is utter simplicity coupled to rigorous order that underlies the phenomenon where it is recognized. Order allows variation but imposes at the same time a canalization that is patent in what we call evolution, being it that of galaxies or of living organisms.

The existence of relationships between the two extreme levels of organization, that of stellar nurseries and of humans, could not be even considered, or searched, if a rigid underlying geometry were not a common denominator of their construction.

Part I

The Skeleton: The Primary Structure that Sets Vertebrates Apart—Is an Evasive Structure

2 Well-Defined Geometric Patterns Appeared in Animal Evolution Independently of the Skeleton

2.1 Regular Color Patterns Arose in Insects Exhibiting the Same Features that Later Appeared in Vertebrates

The vertebrates seem to us such extraordinary creatures, so diversified and so elaborate in many ways, that we tend to think of them as great innovators embodying many novel evolutionary traits. This is true to a large extent, but the vertebrates did not invent everything.

When it comes to the geometric pattern, mainly caused by the regular distribution of pigments, they seem to have invented nothing or little. The transversal and longitudinal parallel stripes that are so distinct in the body of reptiles and mammals were already present in invertebrates. The same is true of the splendid circular markings occurring in fishes and tortoises that were also found in insects. Actually, this ordered pigment distribution occurred with no relation to any internal skeleton—the main trait of the vertebrates.

A few examples illustrate this situation. Parallel stripes located at regular intervals cover the wings of the Colorado beetle. The brown lines are parallel to one another and are regularly distributed along the body axis of this insect, contrasting sharply with a bright yellow background. Stripes distributed at 90° with these ones are found in the wasp *Vespula germanica*. Black bands accompanied by spots are parallel to each other crossing transversally the light yellow abdomen (Fig. 2.1).

Circular spots, that at once draw attention, tend to be composed of several concentric circles of different contrasting colors. They are single, or may be numerous, covering the body, as in the Lantern fly and the Owl butterfly. The name of the butterfly refers directly to a similar pattern seen in birds like owls.

As though concentric circles would not be enough, beautiful colored markings, with the shape of spirals, may also be formed in the wings of mantids (Fig. 2.1). The geometry is extensive.

Not only the insects but other groups of invertebrates display impressive geometric patterns drawn with the help of colourful pigments. In a sea cucumber (Echinoderm) two parallel black bands run along the head-tail axis. These are accompanied by reddish circles that cover the whitish body.

2.2 Spiders Are Also Covered by Exquisite Geometric Patterns that Follow the Same Type of Distribution

There are more species of insects (over one million) than those of all other animals and plants combined. This tremendous variation was accompanied by an impressive rigidity. All insects have six legs. Such a trait could not be altered yet they varied more than any other animal group. The same with the spiders that belong to a different order (Araneae). All have 8 legs and usually 8 eyes, yet have diversified into at least 40,000 species (Burnie 2004).

During the time this differentiation occurred, the spiders, like the insects, covered their bodies with markings that later would appear again in mammals.

A thick longitudinal stripe painted in red, runs along the clear yellow abdomen of the spider *Micromata roseum* dividing it into a left and a right side. A comparable stripe runs along the dorsal rim of the body of some mice species separating the mammalian structure into two regions.

Situated at 90° to the body axis, black transversal lines drawn against a yellow background are seen in the spider *Argiope*. These stripes appear later in zebras with the same orientation and in many other vertebrates.

Parallel dark stripes covering the eight grey legs (*Arctosa cinerea*) are also a feature that later in evolution occurred in the legs of many mammalian species such as hyenas (*Hyaena hyaena*).

A. Lima-de-Faria, *Molecular Origins of Brain and Body Geometry*,
DOI 10.1007/978-3-319-06056-9_2, © Springer International Publishing Switzerland 2014

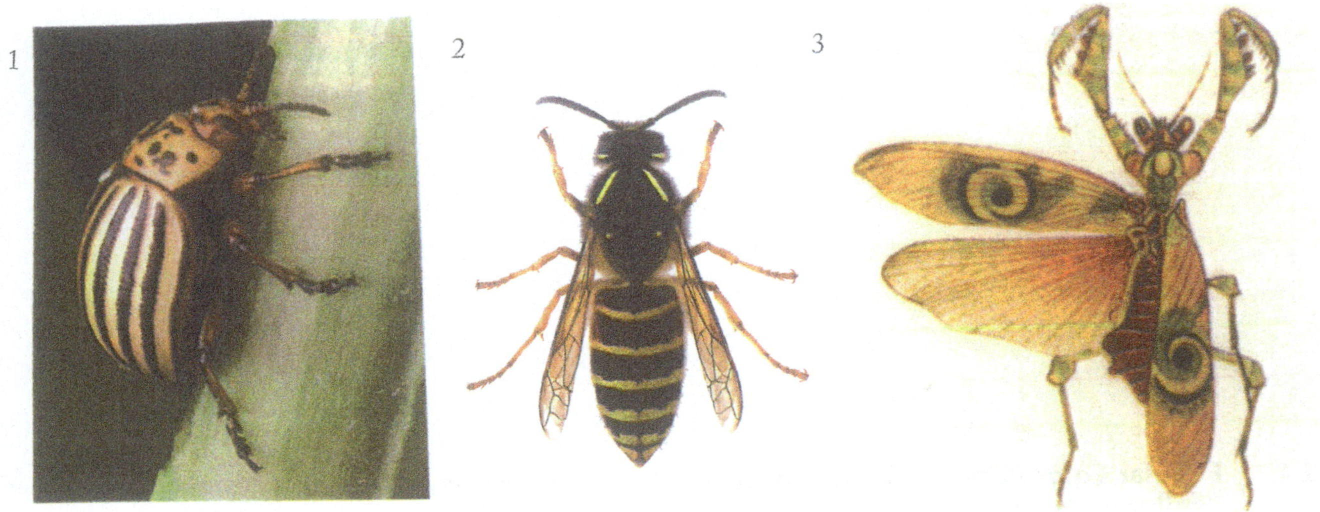

Fig. 2.1 The main patterns observed in vertebrates are already present in invertebrates which had no bony skeleton. (**1**) Longitudinal parallel stripes, Colorado beetle *Leptinotarsa decemilineata*. (**2**) Transversal parallel stripes in the wasp *Dolichovespula norwegica*. (**3**) Spiral (*mainly black*) on wing of insect *Pseudocreobotra*

The most unexpected displays of pattern that are so common in birds and reptiles occurred already at this stage of evolution. Four solid black circles stand up against a red abdomen in *Erescus cinnaberinus*. The name derives from the red colour of cinnabar, also called vermilion red, a compound of mercury that occurs naturally. In tropical snakes black and red are also painted side by side with an equally startling effect.

The genetics of mammals is more extensive than that of invertebrates. The molecular and genetic mechanisms responsible for this geometry are better elucidated at this later level.

Sources of Figures

Fig. 2.1 (**1**) Wikipedia. Author Scott Bauer *USDA*, *ARS*. (**2**) Nationalnyckeln Till Sveriges Flora och Fauna. Steklar: Myror—Getingar. Artdatabanken. Sveriges Lantbruksuniversitet, Sweden 2012 (Fig. page 350). (**3**) Grimaldi, D. and Engel, M.S. 2005. Evolution of the Insects. Cambridge University Press, Cambridge, UK (Fig. 7.90, page 253).

3 Astonishing Gene Permanence Throughout Vertebrates and the Origin of the Skeleton

3.1 Most of the Human Genome Can Be Recognized in Primitive Lancelet Animals: Extreme Gene Conservation Throughout Chordate Evolution that Dates Back Half a Billion Years

Lancelets, or amphioxus, are small worm-like marine animals that share key anatomical and developmental features with vertebrates and tunicates. Together, the vertebrates, tunicates and lancelets constitute the Chordates, having had a common ancestor that lived about 550 million years ago. The lancelets are the modern survivors of ancient chordates.

The structure and gene content of the DNA of the Florida lancelet *Branchiostoma floridae* was analysed in the context of chordate evolution by comparing it with that of tunicates and humans (Putnam et al. 2008). The comparative analysis revealed that 17 ancestral chordate linkage groups, i.e. groups of genes having their location on the same chromosome, are conserved in the modern amphioxus and vertebrate genomes despite the long period of separate evolution. An impressive gene permanence was revealed by the finding that over 90 % of the human genome is encompassed within these linkage groups (Fig. 3.1).

Genome duplications had been considered before to be of general occurrence among vertebrates. Putnam et al. (2008) found two rounds of genome duplications resulting in a fourfold redundancy that is consistent with the whole-genome quadruplication on the vertebrate evolutionary stem.

Despite such drastic reshaping of the chromosome complement the amphioxus-human genetic similarity demonstrates a gene permanence that could not be erased for millions of years.

Jawed vertebrates have four clusters of *Hox* genes each containing up to 14 genes that control the development of the embryo. King et al. (2011) when sequencing the DNA of the shark *Scyliorhinus canicula* and the skate *Leucoraja erinacea* found that the *HoxC* genes were absent. The retention of these four clusters appears to be required for viability because deletion of all copies of *HoxA*, *B*, *C*, or *D* clusters has not been reported in vertebrates. Another fish *Callorhinchus milii* revealed expression of all 45 members of the *HoxA* to *D* clusters. These results disclose once more the tremendous permanence of gene organization and function from fish to humans. Even the position of the two microRNAs families which are embedded between the genes of the four clusters were preserved (Fig. 3.2).

3.2 The Dawn of Mammals: Combination of Reptilian and Mammalian Characters in Platypus Whose DNA Has Been Sequenced

Platypus (*Ornithorhynchus anatinus*) has attracted the attention of zoologists due to its exceptional features. It is a mammal despite its duck-bill and webbed feet. It was placed in a new taxon called the Monotremes because of their common external opening for urogenital and digestive systems. They belong to an ancient line, estimated from fossil and molecular data to be about 166 million years old, that diverged later into the marsupials and the placentals (Warren et al. 2008).

The platypus lays eggs like birds and reptiles but the young depend on milk sucked directly from the abdominal skin, as females lack nipples. There is no scrotum and testes are abdominal. Unique among mammals is their venom delivery system similar to that of reptiles.

The platypus genome has been sequenced and compared to the genomes of other mammals and to that of the chicken. Conservation patterns of microRNAs were compared with those of mammals and the chicken to find out which were shared between these groups. Some of these microRNAs turned out to be unique to Monotremes. Reptile and platypus venom proteins were formed independently but milk protein genes were conserved despite platypuses laying eggs.

A. Lima-de-Faria, *Molecular Origins of Brain and Body Geometry*,
DOI 10.1007/978-3-319-06056-9_3,

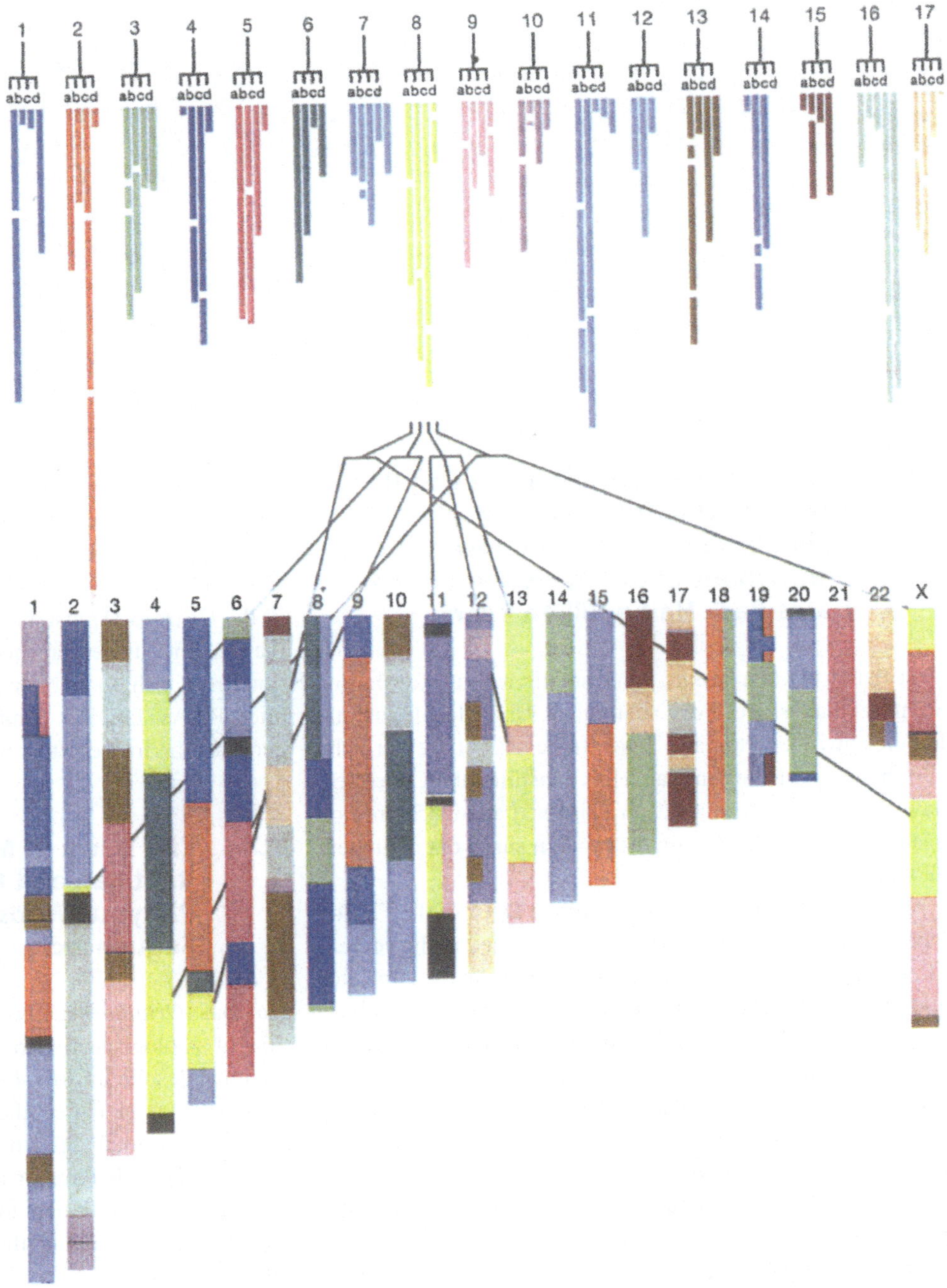

Fig. 3.1 Extreme permanence of gene and chromosome organization. Comparative analysis between the DNA sequences of amphioxus and those of humans. In modern amphioxus, 17 ancestral chordate linkage groups of genes have been conserved, which are present in human chromosomes. Over 90 % of the human genome is encompassed within these linkage groups. The 17 numbers at the *top* represent the 17 ancestral linkage groups. Letters **a–d** represent the four products resulting from two rounds of genome duplication. *Colored bars* are segments of the human genome, shown grouped by ancestral linkage group (*above*) and in context of the human chromosomes (*below*). Over half a billion years of independent evolution separate amphioxus from the human species

3.3 Gene Permanence and Impressive DNA Similarity Allowed the Divergence Between Naked Mole Rats and Humans

The naked mole rat (*Heterocephalus glaber*) is interesting for several reasons. It is a strictly subterranean mammal living in full darkness at low oxygen concentration. Its maximum lifespan exceeds 30 years, making it the longest-living rodent species, and it is cancer resistant.

These features led to an analysis of its DNA in detail both from the cell nucleus and mitochondria. Its comparison with human DNA and that of other species revealed: (1) Transposons (transposable elements of DNA) were 25 % of the genome (in humans 40 %, mouse 37 %, rat 35 %). (2) The number of genes was 22,561 in the mole rat (in humans 22,389, mouse 23,317, rat 22,841). (3) Most of the naked mole rat genome (93 %) showed its gene groups were located at comparable sites in human, mouse and rat with few rearrangements having taken place since these species diverged from a common ancestor. (4) The relationships between these four species were established on the basis of their unique and common gene families (Kim et al. 2011).

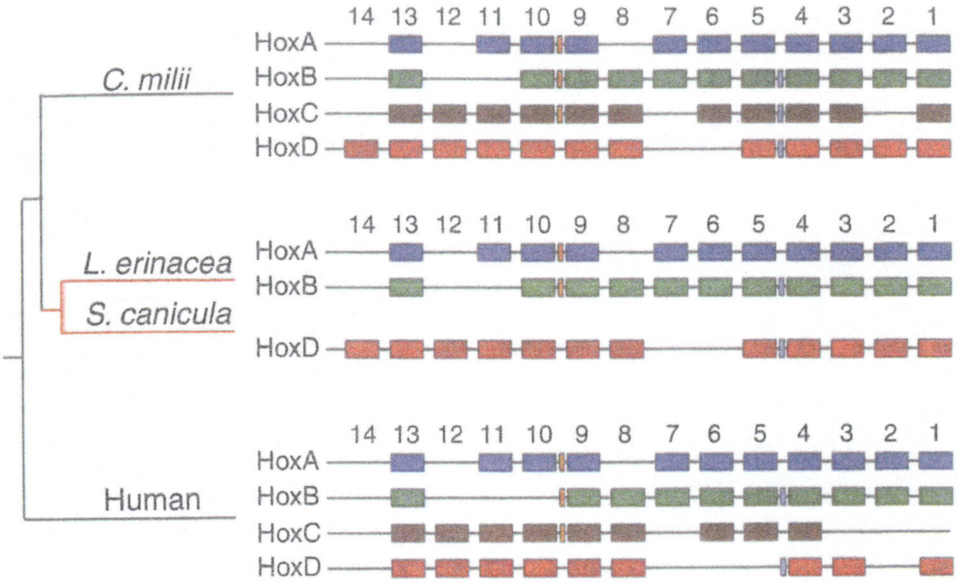

Fig. 3.2 Retention of *Hox* genes in vertebrates. The Elasmobranchs are a sub-class of fishes that include the sharks, skates and rays. The *Hox* genes of humans are compared with those of elasmobranchs, in which a deletion of the *HoxC* cluster occurred, and with those of Holocephalans (*Callorhinchus milii*). Jawed vertebrates, such as humans, typically possess four *Hox* clusters (*HoxA* to *D*, shown in *blue*, *green*, *brown*, and *red*, respectively), each encoding up to 14 *Hox* transcription factors and microRNAs belonging to the *miR-10* (*light blue*) and *miR-196* (*orange*) families. Genomic and transcriptomic sequencing reveals that the skate (*L. erinacea*) and shark (*S. canicula*) retain a full complement of *HoxA*, *B*, and *D* genes but lack *Hox* genes and miRNAs encoded by the *HoxC* cluster. Loss of the *HoxC* cluster (denoted by *red lines*) occurred along the elasmobranch stem

Despite the exceptional traits of the mole rat its DNA analysis disclosed an impressive similarity to humans in: gene number, gene identity and gene location, vindicating earlier findings that demonstrated that the emergence of extreme species differences has been compatible with gene permanence.

3.4 Our Ignorance of the Origin of Humans: The Relationship to Chimpanzees and Gorillas Remains Elusive As Disclosed by DNA Sequencing

The generally accepted view of primate evolution is that chimpanzees and humans share a more recent common ancestor with each other than either of them shares with gorillas. Accordingly, the most closely related sequence for any human gene should be found in the chimpanzee. Humans have many elements of their anatomy and physiology that are similar with both gorillas and chimpanzees. Molecular analysis indicated that humans are closer to African apes than to orangutans and on average closer to chimpanzees than gorillas.

A recent sequencing of the DNA of the Western lowland gorilla (*Gorilla gorilla gorilla*) led to question these assumptions. Genetic similarities among humans and the apes are more complex than expected.

Scally et al. (2012) found that for 70 % of the genomes of the three species, the chimpanzees are more similar to the corresponding sequences in humans than to those in the gorilla, as expected. However, in the remaining 30 % of the genome, gorilla is closer to human or chimpanzee than the later to each other. This information derived mainly from the binding of the protein *CTCF* in human and gorilla. This protein is essential to vertebrate development being involved in transcriptional regulation and protein scaffolding.

A significant question is how these three species acquired their unique morphological, physiological and behavioral characteristics. DNA sequencing is far from explaining this process which needs to be researched at other levels.

3.5 We Are Far from Understanding Mammalian Phylogenetic Trees: Morphological and Molecular Studies Converge in an Attempt to Elucidate Evolution

An exciting transition phase occurs at present in the study of animal evolution. Experts on mammalian systematics are trying to reconcile traditional morphological traits with species comparisons based on DNA analysis. So far the many phylogenetic trees that have been presented must be regarded as preliminary because it turns out that they are not in agreement with each other (Fig. 3.3).

From the fossil record it is possible to determine the interrelationships of the modern species and trace the origin of each major group. But, the collection of fossils has its limitations

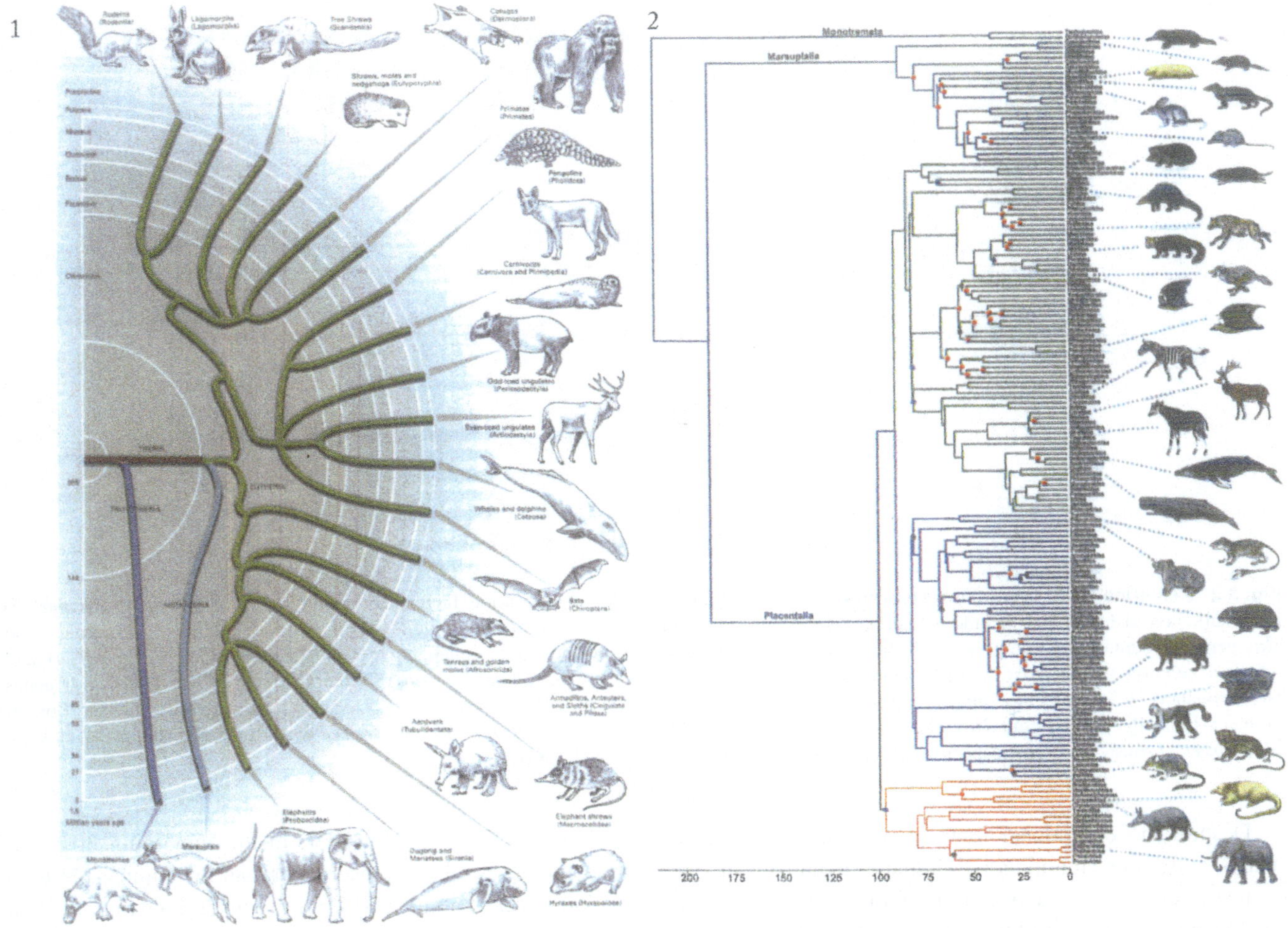

Fig. 3.3 (1) **Evolution of extant groups of mammals and their period of emergence in millions of years**, based on present available evidence according to Wilson and Mittermeier (2009). (2) **Phylogenetic time tree of mammalian families**, created on the basis of an analysis of the amino acid matrix of 164 mammals, rooted with five vertebrate outgroups (chicken, zebrafinch, green anole, frog and zebrafish). As pointed out by the authors (Meredith et al. 2011) there is conflict between amino acid and DNA trees for certain groups and other connections remain difficult to resolve

due to the changing in position of continents with time. Additionally, while some geological periods are represented by a wide spectrum of environments others have a more limited record. At given periods the fossil record remains a patchwork.

One of the main challenges has been the establishment of relationships between species and orders.

Instead of using general similarities, systematists started to concentrate on specialized characters. Soon it turned out that there were serious limitations in this approach. Later the use of phylogenetic trees (also called cladograms) mainly reflected a broad range of anatomical, physiological and behavioral evidence. They represent relationships established on the basis of shared derived characters (Carroll 1988).

Fierce debate issued during a period where many different types of cladograms were published and still continues.

Recently molecular trees, supported mainly by DNA analysis, have disclosed that the relationships established earlier on morphological grounds do not necessarily agree with the new data. One of the reasons is that rates of molecular evolution, far from being constant, show a tremendous variation (Meredith et al. 2011). The result is that molecular phylogenies cannot be dated effectively (Springer et al. 2004; Helgen 2011).

3.6 Contrary to Expectation the Number of Genes Is Not a Good Indicator of Evolution Since It Turns Out to Be Higher in Plants and Invertebrates than in Humans

Most of our notions of superiority based on complexity have turned out to be derived from an antiquated way of thinking. Time and again experimental evidence has disposed of facile interpretations.

One of them is the assumption that the number of genes had to be greater in humans due to their evolutionary position. From the beginning it was considered that humans had to have at least 200,000 genes. As late as 2000 (Gilbert 2000) the figure was 150,000 based on the number of the proteins present in the human body. This value sprang from the one gene–one protein relationship accepted in the 1970s. Soon it became evident that a single gene could give rise to several different proteins and later genes turned out to consist of coding and non-coding regions (exons and introns).

The sequencing of the bases in DNA was supposed to give the ultimate answer. Surprisingly there were about 32,000 (Bork and Copley 2001) genes coding for proteins. Still worse, this figure soon became reduced to circa 20,000 (Pennisi 2003) and it is far from being the final value.

Genes, although well delimited by starting and finishing DNA sequences, have turned out to be of many different types. Moreover, very short sequences of DNA, that before were considered "junk" or "nonsense" DNA (Ohno 1972) turned out to have decisive cell functions.

As plant and protozoa genomes started to be sequenced the picture of complexity related to gene number became even less reliable.

Chlamydomonas reinhardtii is a unicellular alga whose lineage diverged from plants over 1 billion years ago. The sequencing of the base pairs of its DNA revealed 15,143 genes. But the flowering plant *Arabidopsis* had 26,341, exceeding the number of human genes considered at that time to be about 23,000 (Merchant et al. 2007).

Medicago truncatula is a legume plant. A sequence of 94 % of its DNA revealed higher levels of chromosome rearrangements, whole genome duplications and 62,388 genes (Young et al. 2011). Another plant such as the pigeonpea, also a legume food crop, (*Cajanus cajan*) has 48,680 genes as determined by the sequencing of its DNA (Varshney et al. 2012).

To add to this variation are the 18,500 protein coding genes present in the common roundworm *Ascaris suum* (Jex et al. 2011). An equally simple organism is the microscopic crustacean *Daphnia pulex* (water flea) which contains at least 30,907 genes due to an elevated rate of gene duplication (Colbourne et al. 2011).

From these results it is evident that gene number does not contribute directly to elucidate plant and animal evolution, much less that of vertebrates.

3.7 The Number of Chromosomes Is Neither a Reliable Source Since It Varies Drastically in Closely Related Species

The number of chromosomes is known to vary widely from protozoa to humans and within many well-defined groups of living organisms. What is striking, from the point of view of evolution, is that it varies drastically in closely related species belonging to the mammals.

Initially low chromosome numbers were reported in Marsupials. One reasoned that these animals had low numbers because they were primitive mammals which evolved on an isolated continent. Geneticists were baffled when Wuster and Atkin (1972) discovered a species in the deer family having only six chromosomes. A deer is a placental mammal like humans and has the same main type of tissues and organs.

The chromosomes of the Indian or Red Muntjac (*Muntiacus muntjak*) are six in the female and seven in the male body tissues. These chromosomes are quite large and the male has an extra small chromosome called Y_1. But what was even more remarkable was that another Muntjac (*Muntiacus reevesi*) was found to have the same chromosome number as *Homo sapiens* (46). Besides, these chromosomes were quite similar in size and shape to those of the human complement (Hsu and Benirschke 1971). The amounts of DNA in these two Muntjac species and humans were found to be: 5.0×10^9, 5.3×10^9 and 5.6×10^9 nucleotide pairs respectively, which represented a minor variation in DNA content.

Since then there has been a further study of the genus *Muntiacus* which comprises 11 species distributed throughout Asia (Wilson and Mittermeier 2009).

To further bewilder evolutionists another species, also living in the wild, was discovered to have intermediate numbers between 6 and 46. *Muntiacus feae* has a somatic number of 14 (males) and 13 (females).

In addition, the Common Brown Brocket (*Mazama gouazoubira*), which belongs to another genus, but is also a deer (Family Cervidae) has 70 chromosomes (somatic). This animal is, however, most similar to the mentioned species of Muntjacs (Fig. 3.4).

Chromosome numbers are valuable in many respects to understand evolution but, considered separately, they are not reliable guides to find out how species were formed.

The message is clear: we are far from understanding the molecular mechanisms directing evolution.

3.8 The Origin of the Skeleton: The Emergence of the Vertebrates As Well As Their Evolution Are Filled with Question Marks

The main structural feature of the vertebrates is their bony skeleton.

How this main body component has arisen from the structure of invertebrates remains a yet unchartered field of research. No direct relevant genetic information is so far available.

The skeleton was a novelty which emerged in the Cambrian deposits (Carroll 1988). What is still more difficult to explain at present in genetic terms is that once the skeleton appeared it changed its shape and number of components in a dramatic way giving rise to frogs as well as whales. Yet it remained an

Fig. 3.4 Large variation in chromosome number in closely related species within the deer family (Cervidae). (**1**) Red muntjac, *Muntiacus muntjak ssp. montanus*, chromosome number (diploid) in male 7 or 9/in female 6 or 8. (**2**) Reeves's muntjac, *Muntiacus reevesi*, chrom. nr. 46 in both male and female. (**3**) Fea's muntjac, *Muntiacus feae*, chr. nr. 14 (male) 13 (female). (**4**) Common brown brocket, *Mazama gouazoubira*, chrom. nr. in both male and female 70. (**5**) The chromosome complement of *Muntiacus muntjak* male with 7 chromosomes having two Y chromosmes Y_1 and Y_2, and of the female with 6 chromosomes (with two large Xs). (**6**) The chromosome complement of *Muntiacus reevesi* with 46 chromosomes (female). The whole complement, is most similar in number and chromosome size distribution, to that of *Homo sapiens* (diploid number 46)

easily recognizable and well-defined structure in which each vertebra (or other component) finds its homologous part throughout the vertebrates. This is why the main groups of vertebrates can be put side by side without difficulty starting with the ancient fishes and finishing with humans (Fig. 3.5).

As stated by Wellik (2009): "The axial skeleton of all vertebrates is comprised of similar structures that extend from anterior to posterior along the body axis: the occipital skull bones, cervical, thoracic, lumbar, sacral and caudal vertebrae. Despite significant changes in the number and size of these elements during vertebrate evolution, the basic character of these anatomical elements, as well as the order in which they appear, has remained strikingly similar."

3.9 The Skeleton Is Under the Control of *Hox* Genes

Present genetic information reveals that the skeleton of vertebrates is under control of the homeotic genes (also called *Hox* genes). They decide the segmentation of the embryo into compartments that correspond to the anterior, median and posterior sections of the vertebral column.

This information, although highly valuable is still only a step in the understanding of the origin of the skeleton.

A main reason is that the same homeotic genes in humans, and other vertebrates, that decide the sequence of

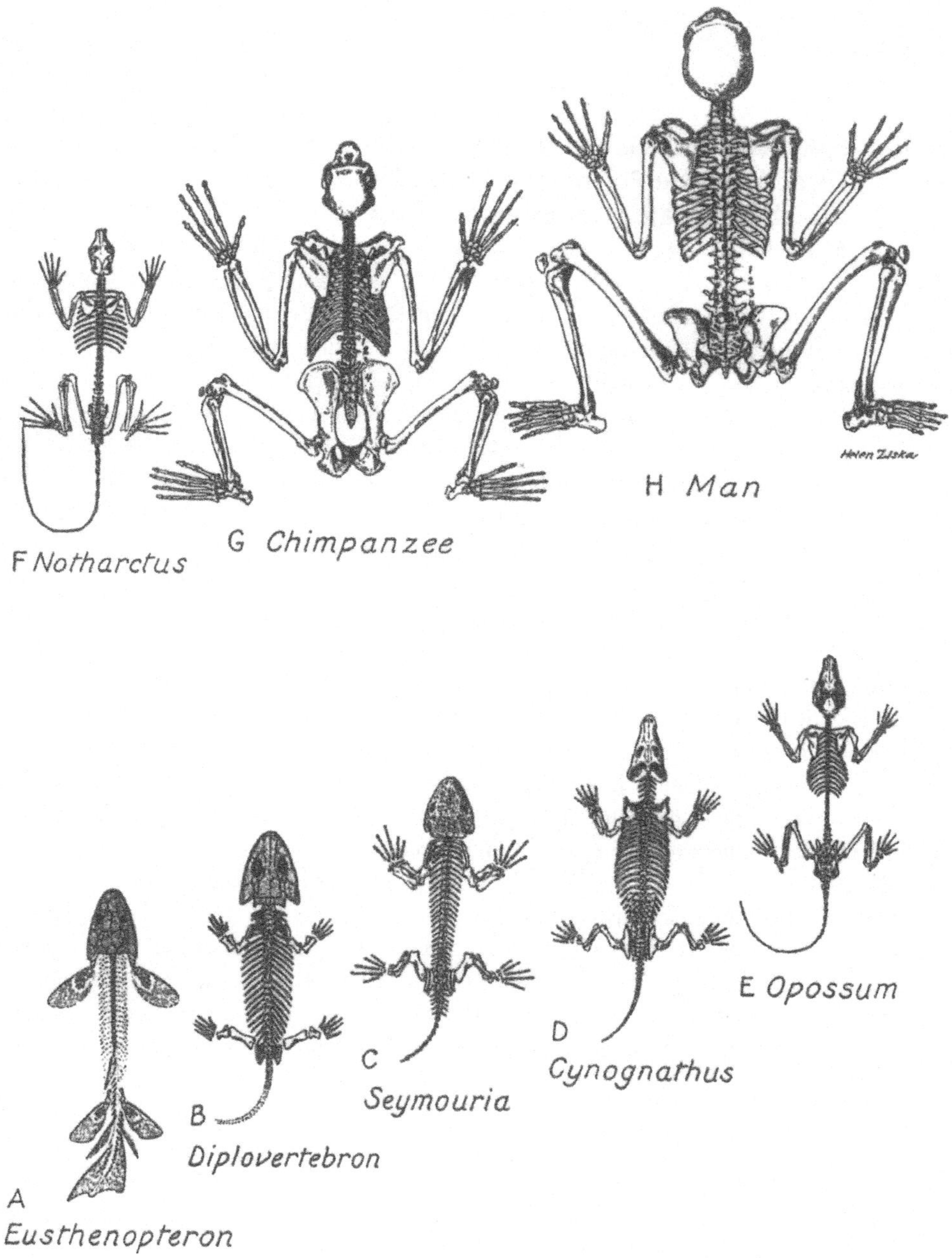

Fig. 3.5 Structural skeletons representative of the evolution of the vertebrates from early fossil tetrapods to humans. *Eusthenopteron* is considered an initial form in which the paired fins were transformed into feet with five digits. Subsequently the limbs got a lateral position

the components of the vertebral column are the same genes that determine the location of the abdomen, thorax and head in insects, which have no internal skeleton. Hence, the fact that the same genes are involved in body segmentation in vertebrates and invertebrates tells us that the formation of the skeleton in itself is not solely the result of their action, although they are part of the process.

It is necessary to realize that the sequential segmentation is a different phenomenon from the emergence of the skeleton. Studies in fish, birds and mice have attempted to elucidate this problem.

Somites are the obvious metameric structures of the vertebrate embryo. They form bilateral pairs flanking the notochord and are created sequentially in an anterior to posterior sequence concomitant with the posterior growth of the trunk and tail. There is a cyclical activation and repression (regulated by a clock) of *notch* genes which determine the formation of somites in zebrafish. This internal clock is also active during chick and mouse embryonic segmentation (Holley 2007).

The combined genetic, molecular and embryologic results confirm the critical role of *Hox* genes in establishing the axial morphology, which lead to what Wellik (2009) called the "Hox code".

3.10 However, *Hox* Genes Do Not Participate Directly in the Formation of the Skull Which Is a Major Component of the Skeleton

The distinction between the action of *Hox* genes and the formation of the entire skeleton was elucidated by studying avian embryos (using grafts of different origin in quail-chick chimaeras). Most of the skull and all the facial skeleton is derived from the cephalic neural crest. This is a group of embryonic cells originating from the ectoderm that migrate to give rise to parts of the nervous system and other cell types. The crest cells participate mainly in the bone and cartilage formation of the anterior region of the skeleton, an area where *Hox* genes are not active (*Hox-negative* domain). Excision from embryos of this neural crest region results in the absence of facial skeleton showing that *Hox* genes cannot replace the bone formation capacity of these cell types. Creuzet et al. (2005) concluded that the formation of the craniofacial skeleton is regulated by an interaction between different types of tissues in which neural crest cells and *Hox* genes collaborate, being active in different areas of the skeleton.

A subsequent approach using the *Dlx* genes in mice confirmed these results. The ancient *Dlx* gene family is represented in vertebrates by six genes *Dlx1* to *Dlx6* all homologous to the *Drosophila* gene called *distal-less* (*dll*). In this fly these genes are crucial for patterning limb and antennae but in the mouse the *Dlx5* and *Dlx6* genes were found to regulate the skeleton. They direct cartilage and bone formation in mammals. Simultaneous mutations of these two genes had a drastic effect. Most of the skeleton of the mouse was formed in the embryo in a regular way but the skull and the tip of the tail were absent. This is a significant finding because it demonstrates that the two extremities of this vertebrate structure are under the control of the same genes, whereas the cervical, thoracic and posterior part of the skeleton have an independent origin (Kraus and Lufkin 2006).

3.11 The Vertebrates Changed from Fins to Limbs: A Skeleton Transformation Not yet Understood in Genetic Terms

Both plants and animals evolved by a shift from water to life on land. Fish are vertebrates which move in water by using fins. The transition to the land led to the appearance of limbed vertebrates, the 4-footed tetrapods, an event that occurred between 420 and 330 million years ago.

Evidence of the evolution of limbed animals from finned precursors comes in many forms but relies mainly on the fossil record. The discovery of the *Tiktaalik* revealed that it had fish-like fin rays which could be used for propulsion in water as well as to support the animal when on land. The skeleton of this species has intermediate characteristics supporting the transitional nature of this vertebrate (Daeschler and Shubin 2011).

This evidence is preliminary. Much needs to be learned about the molecular events that transformed fins into limbs.

Sources of Figures

Fig. 3.1 Putnam, N.H et al. 2008. The amphioxus genome and the evolution of the chordate karyotype. Nature 453: 1064–1072 (Fig. 3, page 1067).

Fig. 3.2 King,B.L. et al. 2011. A natural deletion of the *HoxC* cluster in Elasmobranch fishes. Science 334: 1517 (Fig. 1, page 1517).

Fig. 3.3 (**1**) Wilson, D.E. and Mittermeier, R.A. (Editors) 2009. Handbook of the Mammals of the World, Vol.1. Lynx Edicions, Barcelona, Spain (Vol. 1, Fig. page 45). (**2**) Meredith, R.W. et al. 2011. Impacts of the Cretaceous terrestrial revolution and KPg extinction on mammal diversification. Science 334: 521–524 (Fig. 1, page 522).

Fig. 3.4 Nrs. (**1**) to (**4**) from Wilson, D.E. and Mittermeier, R.A. (Editors) 2011. Handbook of the Mammals of the World, Vol. 2. Lynx Edicions, Barcelona, Spain. (**1**) Vol. 2, Plate 15, Nr. 8; (**2**) Plate 15, Nr. 3; (**3**) Plate 15, Nr. 5; (**4**) Plate 21, Nr. 44. Nr. (**5**) from: Würster, D.H. and Benirschke, K. 1970. Indian muntjac *Muntiacus muntjak*: a deer with a low diploid chromosome number. Science 168: 1364–1366. (**6**) From Lima-de-Faria, A. 1980. Hereditas 93: 47–73 (Page 61).

Fig. 3.5 Gregory, W.K. 1974. Evolution Emerging. Vol. II. Arno Press, New York, U.S.A. (Vol. 2 Fig. 24.2, page 995).

4 Some of the Most Important External Organs in Mammals Have No Skeleton

4.1 External Organs in Mammals, Such As the Penis, May Have Bones or Lack Them, yet Perform Vital Functions

Since vertebrates are characterized by their skeleton, it could be supposed that bones had shaped the totality of their body structure. This is not the case.

During the evolution of mammals several external organs have been formed without an internal bone edifice. It could also be thought that they are of secondary importance. But no, they may be of large dimensions and perform vital functions.

Large ears that can be oriented in different directions have the important function of allowing a better location of sound. They consist of a nearly transparent tissue having no bones, as in the Antelope jackrabbit.

Protruding snouts have no bones that extend to the terminal region of this organ. They are distinct in the Aardvark, Elephant shrew and Proboscis monkey. The name of these last two animals derives from the resemblance of this body part with the elephant's trunk. This large organ is used by the African elephant in many ways (among them courtship) and it can be seen, when compared with the animal's skeleton, to be free of bones (Fig. 4.1).

Mammals derive their name from the mammary glands that are common to all their females. Such an important organ protrudes from the frontal part of the trunk in humans and in other animals is located in the lower ventral region, as in cows. No bones are present in mammary glands.

The penis represents an intermediate situation. In humans it has no bones but in the dog and the otter it contains a bone. Man is an exception, as in most primates a skeletal part is found in their penis called *os penis* or *baculum*. Besides the penis is not only an attribute of males. Females of Spotted hyenas have a long penis which functions as a birth canal and which is devoid of skeleton. At times this narrow opening creates problems during birth (Fig. 4.2).

4.2 The Imposing Humps of Camels and the Large Fins of Whales Have No Skeleton

It is not usually realized that the humps of camels are not part of their skeleton. Both in the dromedary (one-humped) and Bactrian camel (two-humped) this organ is quite large and can be seen, when compared with the skeleton, that it is not part of it. As in the American bison this extension of the body is mainly used for fat storage (Burnie 2004).

Killer whales and dolphins display large dorsal and tail fins which, at times, emerge from the surface of the ocean producing a spectacular sight. The comparison with their skeletons shows that these two organs, so valuable in controlling swimming, are free of bones (Fig. 4.1). In the Bowhead whale there is no dorsal fin but the huge tail fin (over 3 m in width) has a few terminal vertebrae that run only along its median region.

Beavers use their flat tail both as a rudder in swimming and as a fat storage area (Orr 1986; Macdonald 2002). In their skeleton, like in that of the Bowhead whale, only a few terminal vertebrae run into the middle of the broad tail.

The external organs formed without a skeleton, have a particular distribution in the mammalian body. They are located on four main regions: head, frontal and belly region, back of body and tail. They can be distributed on a circle divided into four quadrants each standing for one of the four locations. These organs have basic functions connected with reproduction (penis and mammary gland), sensory perception (ears, snouts, elephant trunk), guidance in swimming (dorsal and tail fins) and fat storage that improves survival (humps and tails).

A. Lima-de-Faria, *Molecular Origins of Brain and Body Geometry*,
DOI 10.1007/978-3-319-06056-9_4, © Springer International Publishing Switzerland 2014

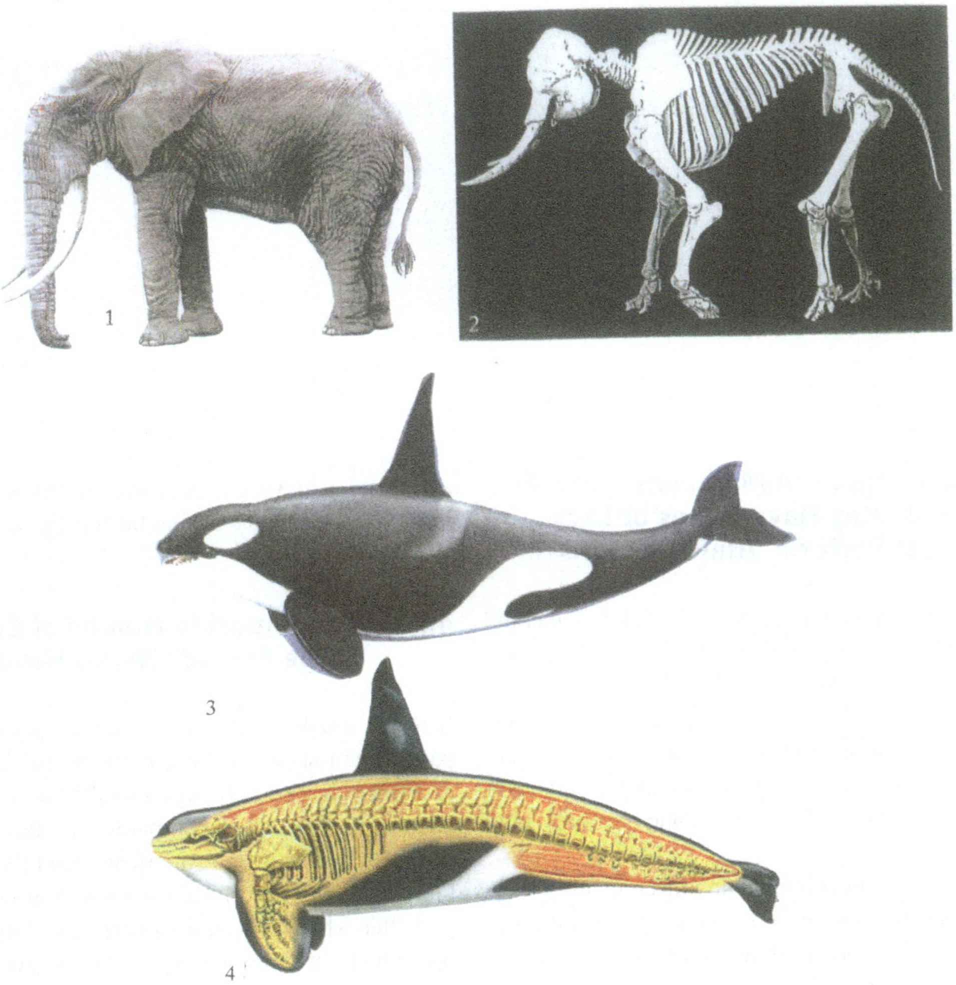

Fig. 4.1 **External organs protruding from the head**. (**1**) African elephant *Loxodonta africana*. (**2**) Skeleton of an elephant showing that the trunk (like the large ear) is an organ that is not part of the skeleton. **Fins project out of the back and build the tail of whales without having a skeleton**. (**3**) Killer whale *Orcinus orca* with its long dorsal fin located on middle of body and a tail fin. (**4**) Skeleton of *Orcinus orca*. The large dorsal fin has no skeleton, neither does the tail

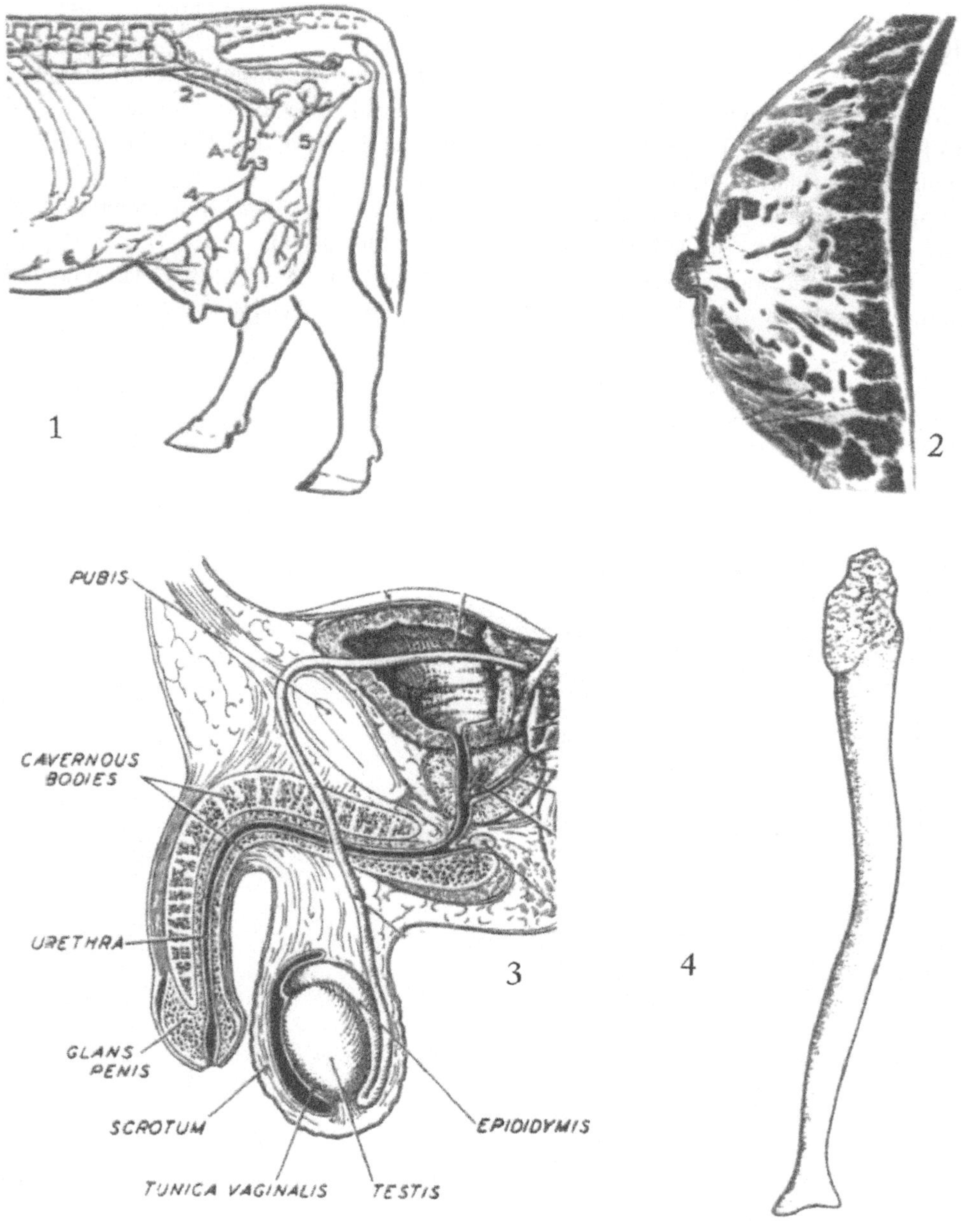

Fig. 4.2 External organs projecting from the frontal part of body: mammary glands, penis and testes. (**1**) Cow *Bos taurus*. Diagram of the arteries which supply the udder with blood, showing two teats. (**2**) Cross section of mammary gland of a pregnant woman with glands secreting milk. (**3**) Cross section of penis and testis of human male. (**4**) The bone (*os penis*) develops in the penis of many mammals and nearly all primates, man being an exception

Sources of Figures

Fig. 4.1 (**1**) Wilson, D.E. and Mittermeier, R.A. (Editors) 2011. Handbook of the Mammals of the World, Vol.2. Lynx Edicions, Barcelona, Spain (Plate 3, Nr. 1). (**2**) File: Elephant Skeleton. jpg—Wikipedia, The Free Encyclopedia (Fig. page 1 of 3). (**3**) Bjärvall, A. and Ullström, S. 1995. Däggdjur. Wahlström and Widstrand, Sweden (Fig. page 262). (**4**) Scientific Illustration, Orca Anatomy. 2012. *Orcinus orca*. Source: www.Carnivoraforum.com (Fig. page 1).

Fig. 4.2 (**1**) Mustafa, A. 2011. Dairy Cattle production. Macdonald, McGill, California, U.S.A. (Detail of Fig. page 9). (**2**) Broman, I. 1933. Människans Anatomi. Gleerups Förlag, Lund, Sweden (Fig. page 170). (**3**) Romer, A.S. and Parsons, T.S. 1978. The Vertebrate Body. Saunders Company, Philadelphia, U.S.A. (Fig. 310, page 300). (**4**) Romer, A.S. and Parsons, T.S. 1978. The Vertebrate Body. Saunders Company, Philadelphia, U.S.A. (Fig. 136, page 154).

5 Body Geometry Follows the Skeleton Only Partially

5.1 A Rhinoceros' Body Configuration Is a Key to the Understanding of the Relationship Between Skeleton and Body Partition

The question then arises: how did the arrival of the internal skeleton affect the coloration of the body in vertebrates?

The vertebral column of a 4-footed animal (tetrapod) has been divided by anatomists into a series of natural regions: cervical, thoracic, lumbar, sacral and caudal (Fig. 5.1). This internal division cannot be recognized by inspecting the exterior of the body of most mammals. A hippopotamus or a white bear, have a body covered by an even skin (or fur) and color that does not show any obvious division into compartments. In them there is no partition of the outer body that corresponds to the anatomic regions of the vertebral column except for the tail which is minute.

There are, however, a number of mammalian species that exhibit a particular relationship between the internal skeleton and the outer body morphology or its coloration.

The Greater One-horned Rhinoceros *Rhinoceros unicornis* has an even grey coloration but it has an external body segmentation which closely follows the parts of the endoskeleton (Fig. 5.1). The external regions are numbered 1–10: skull (1), cervical (2), cervical plus thoracic (3), lumbar (5), sacral (6), caudal (8). In addition the scapula and the humerus encompass region 4, the pelvis and the femur become region 7. The other bones of the forelimbs and hindlimbs are 9 and 10 respectively. These regions which have well-defined limits in the outer body of the Rhinoceros help to understand the distribution of the diverse coloration patterns present in other vertebrates.

5.2 The Mosaic Pattern of the Civet Can Be Divided into 10 Different Regions

Few bodies are such a mosaic of horizontal and vertical stripes as that of the African civet, *Civettictis civetta*. These form 10 distinct regions that can be compared with the skeleton of the common genet *Genetta genetta*. (1) The head has eyes surrounded by large dark circular patches. (2) Nearly horizontal thick stripes cover the neck. (3) The region of the scapula is divided into four oblique and four vertical stripes. (4) The lower thorax has irregular patches. (5) and (6) The lumbar and sacral parts are covered by thick patches that are partly irregular. (7) The pelvis and femur build a series of horizontal lines that run into the hindlegs. (8) Thick transversal stripes are found along the tail. (9) and (10) Both the forelimbs and the hindlimbs are mainly dark. The closely related species, the large-spotted civet, has a pattern that is slightly different, but that can be also divided into ten main regions related to the skeleton (Fig. 5.2).

5.3 The Zebra Is a Lesson in Geometry

Skeletons of zebras are found in museums but have been badly preserved. The result is that good images are difficult to obtain. On the other hand the skeleton of the horse (*Equus caballus*), which is a close relative of the zebra, has been described in detail. The Mountain Zebra *Equus zebra* (subspecies hartmannae) has a pattern that seems to be most complex, but this complexity partly vanishes when it can be seen that the geometric design can be divided into regions that follow closely the internal skeleton. (1) In the head stripes with three different orientations are found in the front and sides as a result of the dominating position of the eyes. (2) The long neck is covered by nearly vertical stripes that are parallel to one another. (3) In the thorax (back vertebrae with ribs) the same pattern prevails. (4) The vertebrae of the loin and of the hip are covered by short vertical stripes. (5) The pelvis and part of the abdominal region are separated from the rest of the trunk by broad white and black bands that instead run in a partly horizontal fashion. (6) and (7) The forelegs and hindlegs display horizontal stripes which cover the limbs from top to bottom (Fig. 5.2).

A. Lima-de-Faria, *Molecular Origins of Brain and Body Geometry*,
DOI 10.1007/978-3-319-06056-9_5, © Springer International Publishing Switzerland 2014

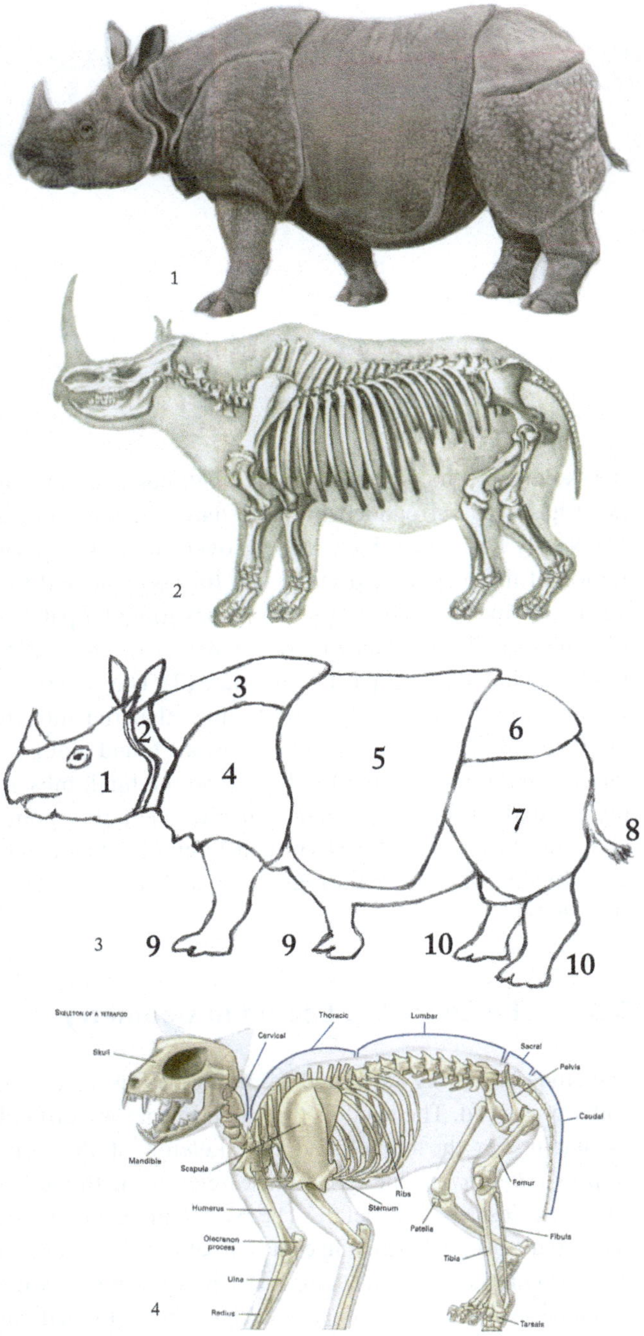

Fig. 5.1 External body segmentation in mammals. (**1**) Greater One-horned Rhinoceros *Rhinoceros unicornis*. (**2**) Skeleton of *Rhinoceros tichorhinus*. (**3**) Numbering of the different body regions in the Greater One-horned Rhinoceros. Part 3 of body includes also neck vertebrae. (**4**) Skeleton of a tetrapod (carnivore) with the natural division of the vertebral column into five sections: cervical, thoracic, lumbar, sacral and caudal

Fig. 5.2 Civets display well-defined body sections. (**1**) Large-spotted civet *Viverra megaspila*. **The apparently complex pattern of zebras is also divisible into segments.** (**2**) Mountain Zebra *Equus zebra* ssp. hartmannae. (**3**) Skeleton of a horse (*Equus caballus*) with its vertebral column divided into: neck, back, loin, hip and tail

Civet and zebra are among the species of mammals where body geometry is most conspicuous. The comparison with their skeletons reveals a relationship that is in a way elusive and at the same time revealing. On one hand there is a division of the pattern that follows the main parts of the skeleton. On the other hand the actual coloration is independent of it because the stripes may have a vertical, oblique or horizontal direction which is not necessarily related to the type of bones lying underneath. Moreover, the same pattern can be repeated at unrelated locations.

5.4 Body Coloration in Relation to the Skeleton in Birds, Reptiles, Amphibians and Fish

The birds are a case apart because their coloration is not written directly on their body but on the long feathers that cover it. The compact skeleton of birds and the presence of wings add to a stronger divorce between the coloration and the distribution of the skeletal bones. There is only a partial agreement when main regions such as the head, neck, thorax, and abdomen are considered. The tail consists of only a few shortened vertebrae. The long feathers, in some species may attain several meters (tail of Japanese fowl). The feathers are responsible for the distribution of the coloration and as a consequence are to a still larger extent independent of the skeleton. Only the beak, legs and feet of birds are usually free of feathers.

Reptiles are like mammals in the sense that their coloration is formed along the skin (on scales and hairs respectively). As such the correspondence with the skeleton can be better analysed. When one compares the green and white broad areas of the body of the Fijian crested iguana with the skeleton of a common lizard, which is similar in its proportions to that of the iguana, the divisions of the coloration partly follow those of the skeleton. The green head is accompanied by a white transversal band that starts at the mouth and reaches the back of the neck. The trunk is divided into three green regions separated by two white vertical bands. This transversal coloration extends along the tail which in the lizard incloses as many as 39 vertebrae.

In most amphibians no tail is present and the skeletal part of the trunk is extremely reduced, only the head and the hindlegs are large. The body coloration of a frog, with its broad transversal bands, across the legs, has only a distant relation with the skeleton.

The same situation is evident in fish where spots cover the main part of the salmon's body.

In conclusion: The skeleton seems to have created avenues for the distribution of the pigmentation making it easier for it to follow given directions, and it also created barriers delimited by the length and number of the various types of bones. At the same time the coloration has known no frontiers. It has traversed one region moving into another without taking the skeleton into consideration.

Sources of Figures

Fig. 5.1 (**1**) Wilson, D.E. and Mittermeier, R.A. (Editors) 2011. Handbook of the Mammals of the World, Vol. 2. Lynx Edicions, Barcelona, Spain (Vol. 2, Plate 6, Nr. 3). (**2**) Brandt, J.F. 1877. Versuch einer Monographie der Tichorhinen Nashörner. Mémoires Académie Imperiale des Sciences, St. Petersbourg, Vol. 24(4) VII Série. (**3**) Original Drawing. (**4**) Wilson, D.E. and Mittermeier, R.A. (Editors) 2009. Handbook of the Mammals of the World, Vol. 1. Lynx Edicions, Barcelona, Spain (Vol. 1, Fig. page 21).

Fig. 5.2 (**1**) Wilson, D.E. and Mittermeier, R.A. (Editors) 2009. Handbook of the Mammals of the World, Vol. 1. Lynx Edicions, Barcelona, Spain (Vol.1, Plate 12, Nr. 4). (**2**) Wilson, D.E. and Mittermeier, R.A. (Editors) 2011. Handbook of the Mammals of the World, Vol. 2. Lynx Edicions, Barcelona, Spain (Vol. 2, Plate 5, Nr. 7). (**3**) Sandars, E. 1937. A Beast Book for the Pocket. Oxford University Press, London, UK (Fig. page 267).

Part II

The Geometry of the Vertebrate Body Takes the Form of Parallel Stripes and Concentric Circles Which Are the Products of Molecular Cascades

6 Longitudinal and Transversal Stripes in Vertebrates

6.1 Tropical Fish, Turtles and Zebras Have Patterns As Exquisite As Those of Birds-of-Paradise

The molecular geometry found in birds is easily recognizable. The feather pigments, with colors covering the whole rainbow, delimit the geometric traits in a precise way. This is why the birds with their exquisite markings have had enchanted observers for generations. The genetic information available shows the pigments to be directed to their locations in the body by proteins and other molecules produced by genes. This is why this geometry was found to be molecularly determined (Lima-de-Faria 2012).

When considering other vertebrates the impression is that they are far from displaying such geometry. Rhinoceroses are all grey, lions are brown, snakes may be all black, and sardines are silvery.

However, an extensive analysis of the body patterns of other vertebrates discloses a geometry even more distinct than that of the birds. Tropical fish turn out to be a kaleidoscope of colors and display impressive patterns that surpass those of the birds. Some reptiles are also so exquisite that they are lessons in geometry and, among the mammals, the zebras have the most unexpected body decorations.

In addition, novel features emerge that better explain the complex configurations found in some species that before could not be understood.

The molecular origin of this geometry can be even more clearly defined because the genetics of humans and mammals is better known than that of birds.

6.2 Parallel Stripes Following the Head–Tail Axis in Vertebrates

In birds, black and white marks replace each other at the same body segment in different species. This was found to be a general feature, that occurred time and again. The black appeared as a negative print of the white.

This phenomenon is also evident in mammals. A longitudinal stripe is most distinct on their back that can be black or white depending on the species. In the Northern birch mouse and the lemur (*Phaner electromontis*) it is black and sharply delimited. It starts at the top of the head and runs along the dorsal rim of the body finishing before the beginning of the tail. In the Striped possum a similar black stripe follows the same path but runs along the tail only finishing near its end. This stripe changes color becoming white in the Large-toothed Ferret-badger. It has an identical distribution to that of the Northern birch mouse but is like a negative print (Fig. 6.1).

Stripes running along the vertebral column and starting at the head are quite rare in birds. This may be due to the compressed shape of their skeleton in which the vertebral column is much shorter than in lizards and mammals. A few birds such as the Crab plover show a broad black band along their dorsal rim.

Amphibians could not be more explicit in the sharpness and color of their dorsal rims. These build well-delimited stripes which are brightly colored. In newts these extend to the tail and in the Mexican toad an interrupted red stripe runs from the head to the rear.

Reptiles and fish do not lag behind. The Common garter snake has a white stripe that goes from the head to the very end of the tail. Fishes also have colored stripes (bright yellow or black) starting at the head and finishing at the caudal fin or extending to its end. These stripes may follow the vertebral column or run above it (Fig. 6.1).

The long dorsal line is not alone. In mammals it becomes accompanied by a series of stripes which are parallel to one another and which are intercalated by white (or yellowish) bands. Moreover, these tend to be spaced by an interval of the same magnitude. They are usually found on both sides of the body but, they do not seem to extend to the ventral region (Fig. 6.2).

The same situation occurs in birds, reptiles, amphibians and fish. In birds multiple parallel stripes are most distinct in the ventral region, whereas in the other groups they are

A. Lima-de-Faria, *Molecular Origins of Brain and Body Geometry*,
DOI 10.1007/978-3-319-06056-9_6,

Fig. 6.1 Longitudinal stripe following the location of the vertebral column. (**1**) Mouse *Sicista subtilis* which has a *black stripe* like *Sicista betulina* (Northern birchmouse). (**2**) Montagne d'Ambre Fork-marked Lemur *Phaner electromontis* with a *black stripe*. (**3**) In this fish species the dorsal stripe does not follow directly the vertebral column (that has a median position) but is located above it following its course (*Lophotus lacepede*). (**4**) Common garter snake *Thamnophis sirtalis*. *White stripe* from head to end of tail

Fig. 6.2 Parallel longitudinal stripes built at regular intervals. (**1**) The mammal Grandidier's Vontsira *Galidictis grandidieri*. Four parallel *black stripes* covering its body. They are spaced at the same distance from one another. (**2**) Lesser Antillean Saltator *Saltator albicollis* with four parallel stripes running from head to vent. (**3**) Leathery Turtle *Dermochelys coriacea*. The shell consists of several plates disposed longitudinally. (**4**) The tropical fish *Sargocentron seychellense*

Fig. 6.3 **Transversal stripes cover most of the body and tend to be parallel and located at regular intervals.** (**1**) Male Zebra duiker *Cephalophus zebra*. (**2**) Bongo *Tragelaphus eurycerus*. (**3**) Green Avadavat *Amandava formosa*. (**4**) Eastern coral snake *Micrurus fulvius*. (**5**) The male of *Genicanthus caudovittatus*

located dorsally and laterally. In amphibians the stripes may extend or be absent from the head and in fish they may be present or absent in the caudal fin. Remarkable is that turtles, which are known for their shells consisting of polyhedrons can build a carapace that consists of longitudinally aligned plates having the same orientation along the body that is present in the stripes of mammals (Fig. 6.2).

6.3 Lines Running at 90° to the Main Body Axis

The longitudinal orientation can be fully reversed. Instead transversal lines build angles of 90° with the main axis. These stripes are black intercalated by white or yellowish ones. They are also spaced at regular intervals and in three species (Numbat, Banded Duiker and Tasmanian "Wolf") they cover most of the central part of the body (being as many as 16) but are absent on the ventral region, head, neck, tail and legs.

The design of the male Bongo is exquisite. Against a brilliant reddish color are drawn pure white thin lines. These are as many as 14 and run down from the dorsal rim like dripping white ink. They are absent from the head, neck, ventral region and legs. Instead on the head, neck and legs appear white marks (at well delimited regions) which build a separate pattern. There are also differences between the front and rear legs (Fig. 6.3).

The tails of some mammalian species display the most impressive transversal bands where black is intercalated with white. The tails of the three mammals: Crested genet, Ringtail and Ring-tailed lemur, could not be more explicit. The broad bands are as many as 15. The breadth of the black bands as well as that of the intercalary white ones is most regular (Fig. 6.4).

The birds and other vertebrate groups have also transversal stripes but these show particularities in every case. In birds, they may have a dorsal position, but usually, the alternating dark and light bands, are ventral (Fig. 6.3). The snakes dispel any doubts about their geometry. Black transversal stripes cover their entire body from head to tip of tail. These are laid against a background of red combined with yellow, and are also spaced at regular intervals indicating an ordered gene action (Fig. 6.3).

Other reptiles, have transversal stripes (black plus white) that cover the body from the head to the tail, and include the legs (Ringtailed gecko). In others the body and tail are covered by dark transversal stripes, the legs may be or may not be included in this configuration.

One would think that the amphibians would be outside such a description. Salamanders have transversal stripes covering the whole body which are as distinct as those of mammals and reptiles.

Fig. 6.4 **Transversal stripes covering tail of mammals.** (**1**) Crested genet *Genetta cristata*. (**2**) Ringtail *Bassariscus astutus*

Fishes paint themselves with transversal bands (Fig. 6.3). In other species a splendid red is covered by not less than eight shining silvery bands. It is not by accident that its name is Harlequin.

6.4 Independence of Band Orientation Whithin the Same Individual

The independence of pattern formation takes different forms. Who could imagine that longitudinal and transversal stripes would be formed, side by side, and in the same part of the body? The mammals, and other vertebrates, leave no doubt about this independence as can be seen in the Western spotted skunk and unrelated species like the Western Bushbuck and the Yellow-striped Chevrotain (Fig. 6.5). Birds, reptiles and fishes have comparable configurations but the amphibians have so far not revealed this particularity. The transversal direction can change into a longitudinal one within the same individual. This can be actually followed as there are intermediate oblique positions (Fig. 6.5). This change of direction, so patent in fish, helps to understand the configuration present in zebras in which the stripes also modify their direction from a horizontal to a vertical position within the same individual.

As described in Chap. 22, fishes have the ability to change their transversal stripes into longitudinal ones within hours, a phenomenon that is most significant for understanding the formation of the body geometry.

Fig. 6.5 Combination of longitudinal and transversal stripes in the same organism. (**1**) Eastern spotted skunk *Spilogale putorius*. (**2**) Bar-crested Antshrike *Thamnophilus multistriatus*. The stripes of the back of the head and neck tend to be longitudinal, whereas those of the breast and belly are transversal. (**3**) *Horizontal stripes* cover the head but change direction in the main body (*Pomacanthus annularis*)

Sources of Figures

Fig. 6.1 (**1**) Bjärvall, A. and Ullström S. 1995. Däggdjur. Wahlström and Widstrand, Sweden (Fig. page 139). (**2**) Mittermeier, R.A. et al. (Editors) 2013. Handbook of the Mammals of the World, Vol. 3. Lynx Edicions, Barcelona, Spain (Vol. 3, Plate 4, Nr. 31). (**3**) Smith, M.M. and Heemstra, P.C. (Editors) 1986. Smiths' Sea Fishes. Springer, Berlin (Fig. Plate 17, Nr. 121.2). (**4**) WIKIMEDIA, org/ WIKIPEDIA/commons.

Fig. 6.2 (**1**) Wilson, D.E. and Mittermeier, R.A. (Editors) 2009. Handbook of the Mammals of the World, Vol. 1. Lynx Edicions, Barcelona, Spain (Vol. 1, Plate 20, Nr. 6). (**2**) del Hoyo, J. et al. (Editors) 2011. Handbook of the Birds of the World, Vol. 16. Lynx Edicions, Barcelona, Spain (Vol. 16, Plate 37, Nr. 30). (**3**) Cabrera, A. (Editor) 1935. Historia Natural. Instituto Gallach, Barcelona, Spain (Fig. page 392). (**4**) Smith, M.M. and Heemstra, P.C. (Editors) 1986. Smiths' Sea Fishes. Springer, Berlin (Fig. Plate 21, Nr. 132.15).

Fig. 6.3 (**1**) and (**2**) Wilson, D.E. and Mittermeier, R.A. (Editors). 2011. Handbook of the Mammals of the World, Vol. 2. Lynx Edicions, Barcelona, Spain (Vol. 2, Plate 27, Nr. 31 and Plate 52, Nr. 243). (**3**) del Hoyo, J. et al. (Editors) 2010. Handbook of the Birds of the World, Vol. 15. Lynx Edicions, Barcelona, Spain (Vol. 15, Plate 14, Nr. 1). (**4**) WIKIMEDIA. org/WIKIPEDIAcommons. (**5**) Smith, M.M. and Heemstra, P.C. (Editors) 1986. Smiths' Sea Fishes. Springer, Berlin [Fig. Plate 72, Nr. 204.6(A)].

Fig. 6.4 (**1**) Wilson, D.E. and Mittermeier, R.A. (Editors) 2009. Handbook of the Mammals of the World, Vol. 1. Lynx Edicions, Barcelona, Spain (Vol. 1, Plate 13, Nr. 12), (**2**) Wilson, D.E. and Mittermeier, R.A. (Editors) 2009. Handbook of the Mammals of the World, Vol. 1. Lynx Edicions, Barcelona, Spain (Vol. 1, Plate 30, Nr. 4).

Fig. 6.5 (**1**) Wilson, D.E. and Mittermeier, R.A. (Editors) 2009. Handbook of the Mammals of the World, Vol. 1. Lynx Edicions, Barcelona, Spain (Vol. 1, Plate 31, Nr. 10). (**2**) del Hoyo, J. et al. (Editors) 2003. Handbook of the Birds of the World, Vol. 8. Lynx Edicions, Barcelona, Spain (Vol. 8, Plate 42, Nr. 19). (**3**) Smith, M.M. and Heemstra, P.C. (Editors) 1986. Smith's Sea Fishes. Springer, Berlin [Fig. Plate 73, Nr. 204.7(A)].

7 The Geometric Patterns of Zebras and of Transgenic Mice Are Directed by Specific Genes: The Geometry of Mammals Becomes Also Molecular

7.1 Earlier Interpretations of the Zebra's Configuration

No other African animal displays such an attractive body geometry than the zebra. It is so distinct that it cannot be easily confused with any other species roaming on the savanna. Yet it appears as highly complex. Black stripes alternate with white ones that tend to be parallel to each other. Surprisingly they change their orientation by as much as 90 degrees within the same individual. They may have a longitudinal or a transversal orientation along the body axis and in some regions they occupy an intermediate direction being oblique.

Biologists have produced several interpretations of these conspicuous markings (Macdonald 2002). From the beginning they were considered a form of camouflage, but this idea was soon abandoned because zebras gather in great numbers, at exposed sites, and do not hide from predators. Secondly, it was claimed that stripes evolved to deter harmful flies, but this disregarded the fact that insects are no hazard in most zebra habitats. Thirdly, they have been considered to have a social function promoting group cohesion as stripes would stimulate grooming. Fourthly, the ancestors of all equines (which comprise horses, zebras and asses) were striped, but selection in cold climates led them to develop a homogenous coloration. The second interpretation has been recently revived. Polarized light was employed to show that a zebra-striped pattern attracted fewer horseflies than homogenous colors such as black, grey or white (Egri et al. 2012).

These hypotheses have been valuable because they incited other scientists to carry out further research.

The elucidation of the zebra configuration demands the coherent assembly of the following information: (1) Detailed comparison of the pattern found in all species and subspecies. (2) A comparison with recently extinct species. (3) Crosses between different zebras. (4) Analysis of zebra DNA. (5) Genetic analysis of the crosses performed between zebras and asses (which are uniformly colored). (6) Experiments on chimaeric mice which display a pattern comparable to that of zebras. (7) The knowledge of genes involved in the body coloration of mammals. (8) The transformation of longitudinal stripes into transversal ones occurring spontaneously in fishes within the same individual.

When these data are put together the complexity that has dazzled ecologists is minimized and the zebra pattern becomes intelligible.

7.2 Genetic Evidence for the Occurrence of Three Separate Body Patterns in Zebras As a Result of Crosses Between Donkeys and Zebras

The stripes shown by zebras appear also in other related species but in lesser numbers and a simpler form. This means that they do not come from nowhere, but have precedents.

The Tibetan Wild Ass *Equus kiang* has a uniform coloration without stripes, but the Wild Ass *Equus asinus* shows longitudinal stripes on the lower part of the forelegs and hindlegs which finish near the knee. The asses are closely related to zebras since they belong to the same genus (Fig. 7.1).

In the Okapi, *Okapia johnstoni*, which is an Ungulate like the zebras the stripes occupy instead the upper part of the legs (both forelimbs and hindlimbs) and extend to the rump finishing where the tail begins maintaining all the way a horizontal position (Fig. 7.1).

An animal which is a key to the understanding of the zebra pattern was obtained in captivity by the "Quagga Project" in South Africa (Fig. 7.1). Crosses were made between Plains zebras (comprising 6 subspecies) which showed reduced stripes over the rear body. After three generations the pattern closely resembled that of the extinct Quagga known from zoos and museums. Analysis of mitochondrial and nuclear DNA of Plains zebra, and DNA of museum specimens, demonstrated that the variation among living and extinct forms is no greater than among modern breeds of horses (Wilson and Mittermeier 2011, Leonard et al. 2005). This animal has clear legs like those of the Tibetan Wild Ass but the frontal half of the body (head, neck and thorax region) is covered by transversal stripes like

A. Lima-de-Faria, *Molecular Origins of Brain and Body Geometry*,
DOI 10.1007/978-3-319-06056-9_7,

Fig. 7.1 The stripe configuration of zebras has been regarded as highly complex and difficult to decipher. It turns out to be composed of three separate patterns which integrate into one another. These are: (1) longitudinal stripes in forelimbs and hindlimbs (2) transversal stripes in head, neck and front of body (3) longitudinal or oblique stripes in rump and back. **(1)** Tibetan Wild Ass *Equus kiang*, has a uniform coloration without stripes. **(2)** Wild Ass *Equus asinus* (*somalicus*) shows longitudinal stripes on lower part of legs. **(3)** Okapi, *Okapia johnstoni*. It has large longitudinal stripes on upper part of legs and rump. **(4)** Experimentally obtained *Equus quagga*. It has no stripes on legs and rump but transversal stripes on middle of trunk, neck and head. Irregular *black lines* are seen in the posterior region because this is a hybrid recently obtained in captivity (compare with Fig. 7.2). Some *black lines* on legs are grass stems (photograph). **(5)** Chapman's zebra *Equus burchelli antiquorum*. A distinct line runs vertically across the middle of the trunk separating sharply the oblique pattern of the rear from the frontal one that is mainly transversal. This species is also called Plains zebra

Fig. 7.2 **Experimental demonstration of the independence of the three patterns found in zebras. Hybrids obtained by crossing donkeys with zebras.** (**1**) Hybrid between a male Grant's zebra with its complete typical pattern and a female black Arabian ass (Colchester Zoo, England; Syndication International, London). Stripes occur only on the legs, the rest of the body is free from pattern. (**2**) Recently extinct Quagga *Equus quagga quagga*. The only Quagga to ever have been photographed alive (Regent's Park Zoo, London, 1870). This species had stripes only in the head, neck and frontal part of trunk (compare with living Plains zebra *Equus quagga* which is similar to the extinct species (Fig. 7.1). (**3**) An accurate painting of the living Quagga made by Nicolas Maréchal in 1793 which is part of the collection of the French National Museum of Natural History

that of a zebra. The regular pattern stops here, the back part of the trunk has an irregular design and the rump is free of markings like the legs (Fig. 7.1). There are, however, photographs and accurate paintings of the extinct Quagga which show that the stripe pattern was solely limited to the frontal part of the body (Fig. 7.2).

Plains zebra (*Equus burchelli* subspecies *chapmanni*) display a critical feature. A distinct vertical line runs across the middle of the trunk separating the transversal stripes that cover part of the head, neck and thorax region, from the oblique stripes that are seen on the rump and back part of the trunk (Fig. 7.1). To be noted is that the forelegs do not follow the pattern of the front part of the body but have the same longitudinal stripes as the hindlimbs. This is the situation found in the Okapi and the wild ass (*E. africanus* also called *E. asinus*).

Hence, three types of patterns, which seem to be independent of each other, are present in these species: (1) The sequence of longitudinal stripes in both forelimbs and hindlimbs (Wild Ass, Okapi and all zebra species). (2) The sequence of transversal stripes in the head and frontal half of body (all zebra species). (3) The sequence of longitudinal or oblique stripes in the back half of the body and rump (most species and subspecies). In all cases, independently of orientation, the stripes tend to be parallel to each other and to be spaced at regular intervals.

During the 1960s there was an upsurge of research on the chromosomes of mammals. Large Atlases were published describing their chomosomes and crosses were made between different species to elucidate genetic differences. This project led to crossing zebras with donkeys using one species either as father or mother. A typical female Grevy zebra *Equus grevyi*, with the whole body covered by stripes, was the mother of a hybrid which had as father a male donkey *Equus asinus* of homogenous grey colour. The resulting "Zebronkey", turned out to have stripes only on the legs (extending a little into the rump like in the Okapi). These were longitudinally oriented, and parallel to each other, as in the mother (Benirschke et al. 1964, Gray 1972).

Another hybrid was obtained between a male Grant's zebra (with its complete typical pattern) and a female black Arabian ass. Again longitudinal stripes occurred only on the four legs, the rest of the body being free of pattern (inclusive the rump) (Fig. 7.2). These hybrids show clearly that the leg pattern is a process independent of the rest of the body.

The independence of the front from the rear pattern is made evident by the existence of an extinct animal which was photographed for the last time in 1870 in London's zoo. This species called Quagga (*Equus quagga quagga*) had legs as well as the rear half of the trunk free of stripes. Only the head, neck and thoracic region had the typical transversal stripes of a zebra. Hence, the frontal body pattern occurred independently of the rear one. These two body patterns actually confront each other and build a sharp frontier between their separate territories. This is evidenced by the formation of the vertical line that separates them in Plains zebra also called Chapman's zebra (Fig. 7.1).

7.3 Parallel Stripes Can Become Concentric Circles and Can Rotate 90°

It ought to be pointed out that the anterior and posterior regions have some particular traits.

In the head, stripes starting on the nose follow the main body axis, but as they reach the eye they do not cross it, as it occurs in some species of birds, but circumvent the eye, changing their direction to become transversal.

A particular feature occurs on the rump of Grevy's zebra *Equus grevyi*. It has a true circular pattern. The region where the tail starts is the center of concentric circles that are divided into two halves by the black stripe that runs along the dorsal rim (Fig. 11.4). Still more remarkable is that a comparable circularity exists in this species at the opposite side of the body. This is found around the neck, which consists of several concentric thick bands with increasing diameter (Fig. 11.2). This double circularity appears as a secondary pattern imposed on the three primary patterns just described.

Perhaps the most fundamental characteristics of the body stripes is both their capacity to rotate by 90 degrees having either a longitudinal or a transversal position, and the fact that they can be seen, within the same animal, to take the intermediate oblique position, apparently moving between these two directions. This reverse of direction within the same animal occurs in an equally distinct way in the Meyer's butterflyfish, an animal without any immediate phylogenetic relation to a zebra. This similarity is in agreement with the fact that many genes are common to all vertebrates remaining unaltered in their function throughout millions of years (Chap. 3).

As it will be seen later (Chap. 22) in fishes the longitudinal stripes can be transformed into transversal ones within the same individual and within hours.

The main geometric features of the body of zebras are represented diagrammatically in Fig. 7.3.

7.4 Chimaeric and Transgenic Mice Elucidate the Zebra Pattern

The genetic events that guided embryonic development remained for many years a black box. Some of the first crucial experiments in this direction were carried out on mice.

Instead of crossing different strains with opposite traits, one resorted to a novel technique. DNA was marked with radioisotopes to follow accurately the fate of individual cells during development. Besides, mice fertilized eggs were flushed out of the uterus of one female and fused in a petri dish with fertilized eggs from another female which carried different genetic markers such as coat color.

The fused eggs which were at different stages of cell segmentation combined easily into a single embryo. This was then replaced into the uterus of one of the females and allowed to develop normally.

The result was startling. To start with, normal mice were born, similar in function and structure to the parents. Secondly, the combination in test tube of fertilized eggs from a black coat with a white coat donor did not result in a homogenous or spotted black, white or grey coat, but in a mouse that looked like a zebra having most of its body covered with broad transversal stripes. A significant detail was that these stripes extended to the tail but were absent from the head and legs (Fig. 7.4) (Mintz and Illmensee 1975, Mintz 1978).

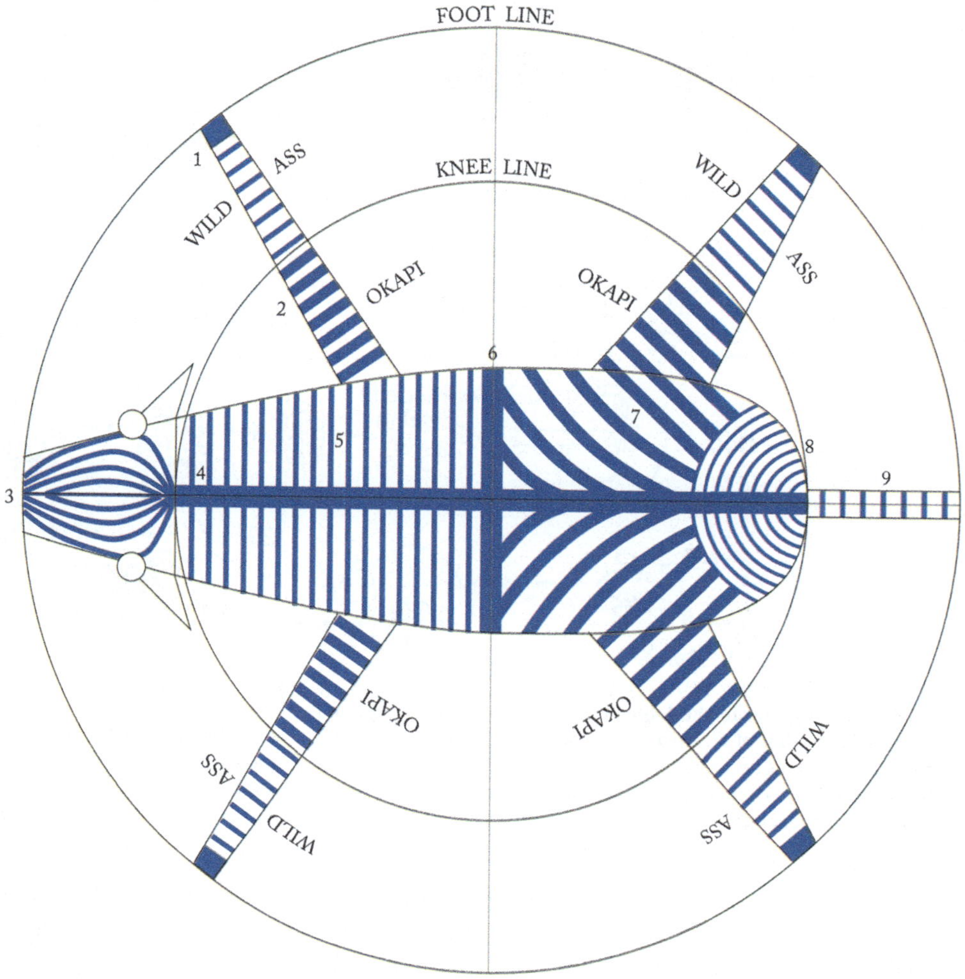

Fig. 7.3 The ordered pattern of a zebra is a chimera combining several independent traits. It is here represented in a geometric form as a result of its highly regular construction. (1) The longitudinally oriented stripes of the lower part of the forelegs and hindlegs are found in the wild ass from the foot to the knee. (2) Thicker stripes, with the same orientation, are present in the Okapi above the knee. In zebras the hindlimbs are usually broader than the forelimbs. This compound leg pattern is the sole trait that appears in hybrids of crosses between zebras and donkeys. (3) Head of Plains zebra seen from above with curved lines following the location of the eyes. (4) A *thick black line* runs above the vertebral column, from head to rump. It is most distinct in Grevy's zebra. (5) A series of transversal stripes cover the sides of the head, neck and frontal part of body. They are the sole pattern found in extinct Quagga. (6) Distinct vertical line dividing the frontal part of the body from the posterior one. This is best seen in Plains zebra. (7) Oblique to longitudinal stripes cover the rear of the body, a feature present in several species. (8) Concentric circles originating at the base of the tail are characteristic of Grevy's zebra. (9) The tail in some species has also transversal stripes

This initial result was followed by other experiments which were addressing other problems such as the origin of cancer cells and the production of transgenic animals. They were carried out in different laboratories but the result was the same—mice with a "zebra coat" instead of a homogenous color or a pattern consisting of black and white spots. Although the cells could be followed with the help of radioisotopes there has been no simple genetic explanation to account for the formation of this unexpected zebra pattern.

7.5 Genes Responsible for Color Variation in Mammals Have Been DNA Sequenced and Explain the Alternating Occurrence of White and Black Stripes

A vast literature has been assembled on genes responsible for coloration in mammals. The molecular analysis has gone as far as to sequence their DNAs.

Mouse DNA of the entire coding region of the *melanocortin-1 receptor gene* (*Mclr*), was sequenced, since a protein-coupled receptor plays a key role in melanogenesis by switching between the production of dark and light pigments (Hoekstra et al. 2006).

This gene which has about 1,000 nucleotides, was found to produce a mutation involving a single nucleotide (thymine to cytosine) that is responsible for change in coat pigmentation. The resulting protein has a single amino acid modification (arginine to cysteine) that leads to radical consequences. Light mice have a version of the gene and all dark mice have the other.

Mclr has an additional property. It acts as a switch in determining the type of pigment a melanocyte produces. When this gene is turned on it signals, through a cascade of effects in the melanocyte cell, for dark melanin to be produced, but when it is turned off light melanin is formed (Hoekstra 2011). White or black mice are born depending on the state of the same gene which may give a high or a low signal to pigment cells.

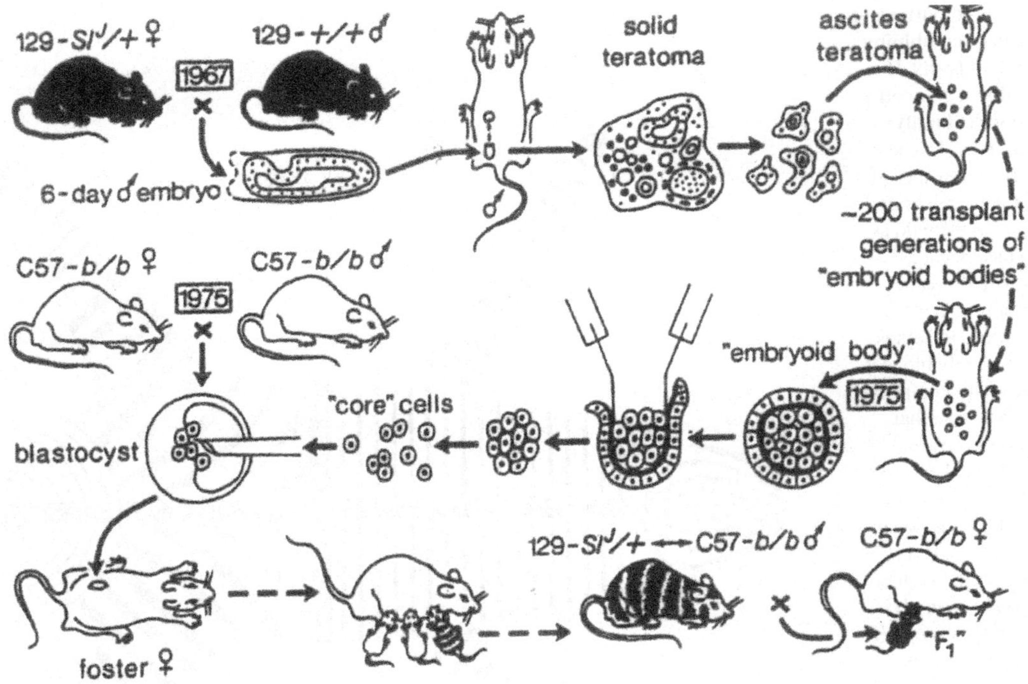

Fig. 7.4 Production in the laboratory of mice with a stripe pattern similar to that of a zebra. Two parent mice, which were *black*, donated embryonal carcinoma cells (6-day embryo). These were injected into blastocysts bearing many genetic markers. After almost 200 transplant generations of "embryoid bodies" the "core" cells were injected into a blastocyst provenient from two *white* mice parents and this was implanted into a foster *white* female. This last animal produced a progeny consisting of white mice and mice with *black and white* transversal stripes, instead of having a homogeneous *grey colour*

Such a simple mechanism may be the source of the occurrence of alternating white and black stripes in zebras if we assume that during embryonic development the *Mclr* gene oscillates with a constant frequency between an activator and a suppressor state. This oscillation would lead to forming black and white stripes which succeed each other in a regular fashion and at equally distant intervals.

A general feature of mammalian patterns is that the same body region, such as the color of the dorsal rim, may be white in one species and black in another. They actually function as one being the "negative" of the other. The state of the *Mclr* gene is also known to change from subspecies to subspecies (Hoekstra et al. 2006) which would account for the sudden emergence of an opposing color at the same body site.

Sources of Figures

Fig. 7.1 (**1**) and (**2**) Wilson, D.E. and Mittermeier, R.A. (Editors) 2011. Handbook of the Mammals of the World, Vol. 2. Lynx Edicions, Barcelona, Spain (Vol. 2, Plate 5, Nr. 3 and Nr. 4). (**3**) Wilson, D.E. and Mittermeier, R.A. (Editors) 2011. Handbook of the Mammals of the World, Vol. 2. Lynx Edicions, Barcelona, Spain (Vol. 2, Plate 56, Nr. 2). (**4**) Wilson, D.E. and Mittermeier, R.A. (Editors) 2011. Handbook of the Mammals of the World, Vol. 2. Lynx Edicions, Barcelona, Spain (Vol. 2, photograph page 137). (**5**) Wilson, D.E. and Mittermeier, R.A. (Editors) 2011. Handbook of the Mammals of the World, Vol. 2. Lynx Edicions, Barcelona, Spain (Vol. 2, Plate 5, Nr. 5).

Fig. 7.2 (**1**) Gray, A.P. 1972. Mammalian Hybrids. Commonwealth Agricultural Bureaux, Farnham Royal, Slough SL2 3BN, UK. Photograph on front page, Syndication International, London, UK. (**2**) Extinct Quagga, Photograph taken at Regent's Park Zoo, London, UK in 1870 (copyright has expired). (**3**) Grassé, P.P. 1977. Larousse Animal Portraits. Hamlyn, London, UK (Fig. facing page 22. Parchment in the French National Museum of Natural History, Paris, painted by Nicolas Maréchal in 1793, Vol. 72, Nr. 93).

Fig. 7.3 Original diagram. Not to be reproduced without the complete reference to this work.

Fig. 7.4 Mintz, B. and Illmensee, K. 1975. Normal genetically mosaic mice produced from malignant teratocarcinoma cells. Proc. Nat. Acad. Sci. U.S.A. 72: 3585–3589 (Fig. 1, page 3586).

8 The Eye: A Main Center of Circularity with Implications for Development and Evolution

8.1 The Eye Arose Before the Brain

We tend to think that the brain, which is the main organ directing our functions and behavior, ought to have been the primary tissue occurring during animal development and evolution.

Plants do not have a brain, yet they develop and confront the environment in a way that in many ways surpasses animals. Already in a tree embryo, its tissues can be seen to separate into those that will later absorb water and minerals from the soil and those that will move in the direction of the sun, obtaining the light energy used in photosynthesis. This growth in opposite directions is controlled by hormones that the plant produces. Plants are self-sufficient, synthesizing the necessary compounds by absorbing the atoms and molecules from the mineral world and from the atmosphere. The ability of plants to extract maximal advantage from the environment is achieved without a brain and is so efficient that it allows them to have the longest life span known among living organisms. Specimens of yew (*Taxus*) may reach an age of 4,000–5,000 years. *Sequoia gigantea* of California and *Pinus aristata* can become 4,000–5,000 years old as authenticated by counts of annual rings (Denffer et al. 1971).

The strongest evidence that the eye arose before the brain is found in single cell organisms, the protists. Dinoflagellates, like *Erythropsis*, have eye organelles consisting of a lens and pigment. They also have retina-like structures with membrane stacks, closely resembling the eyes of multicellular organisms. In *Erythropsidinium* the ocellus consists of: (1) a lens, (2) a fluid filled chamber and (3) a cup covered by light sensitive pigment. This structure is already adaptable and mobile: (1) The lens can change its shape. (2) The pigment cup can move freely in relation to the lens. (3) The whole ocellus can protrude from the cell, thereby allowing it to point in different directions. It is thus not surprising that the dinoflagellates use their eyes to detect the presence of prey just as later invertebrates and vertebrates are able to do (Friday and Ingram 1985). The protozoa have nervous fibrils connected with locomotion but have no brain (Kudo 1971).

As the invertebrates emerged the same situation persisted. Jellyfish, like *Cladonema* or *Chironex*, have highly evolved eyes but lack a brain. The eyes are located at the base of the tentacles and transmit their information directly to the muscles without processing by any intermediate organ. There are a few interneurons and a ring nerve around the belly margin but no real brain. The Box jellyfish *Tripedalia cystophora* has 24 eyes of four different types which it uses to control swim speed and direction (Petie 2012). Gehring (2002) concludes, from his genetic studies on the eyes of mammals and insects, that "the eye as a sensory organ evolved before the brain which is an information-processing organ".

8.2 Instead of Having Eyes Sea Urchins See with Most of Their Body: DNA Sequencing Located Two Photoreceptor Genes

The nervous system of echinoderms, like sea urchins, consists of a primitive nerve ring accompanied by radial nerves. Like plants, sea urchins react to light even though they do not have eyes. DNA sequencing of the sea urchin *Strongylocentrotus purpuratus* localized two genes (*Sp-opsin4* and *Sp-pax6*) which are essential for photoreceptor function and code for the widely occurring eye protein, *opsin*. Photoreceptors are located all over the urchin's surface, the entire adult sea urchin seeing with most of its body (Ullrich-Lüter et al. 2011).

A tree or any other plant, with its many leaves, sees also with most of its body. A careful study of plant leaves using the electron microscope has revealed that they are mosaics

A. Lima-de-Faria, *Molecular Origins of Brain and Body Geometry*,
DOI 10.1007/978-3-319-06056-9_8,

of microlenses, being actually carpets of light-sensitive cells which do not appreciably differ from those of protozoa and invertebrates. In *Fittonia* and *Medicago* plants, the cells of the leaf epidermis are convex having a biconvex lens located at their tip. Their cytoplasm is transparent and lacks pigment allowing the transmission of light to lower palisade tissue, where the chloroplasts, with their chlorophyll pigment receive light and transform it into chemical energy (Vogelmann et al. 1989; Martin et al. 1989).

Sea urchins use their ability to sense light to move in a specific direction. It is well-known that leaves also move in the direction of light, a phenomenon long ago called phototropism. Both types of organisms, without having brain or eyes, see light and behave accordingly.

8.3 The Evolution of the Eye Has Been Directed by Molecular Cascades

Initially, the evolution of the eye was investigated on the basis of morphological and physiological data but lately it has been submitted to molecular analysis. Evolution has led to at least three basically different types of eyes: (1) The camera-type eye with a single lens projecting onto a retina (vertebrates and cephalopods). (2) The compound eye with multiple ommatidia each consisting of a set of receptor cells and a lens of its own (insects and other arthropods). (3) The mirror eye which uses both a lens and a parabolic mirror which project light onto a proximal and a distal retina respectively (*Pecten*, a scallop).

Zoologists originally proposed that the vertebrate and the compound eye of an insect were independent evolutionary events but the comparison of the action of the *Pax6* gene in mammals and insects led Gehring (2002) to conclude that these three types of eyes had a common origin. Molecular analysis of the opsin phototransduction cascade in jellyfish also implied a common origin for vertebrate and invertebrate eyes (Koyanagi et al. 2008).

8.4 Eyes Can Migrate Along the Body and There Are Fishes with Four Eyes

Eyes tend to be located on the frontal region of the head in living organisms but this position is far from being general. By manipulation of genes in *Drosophila* it has been possible to obtain flies with eyes on the legs, on antennae and even on wings (Halder et al. 1995) (Fig. 8.4).

Ancestral vertebrates had a third eye, besides the two lateral eyes, medially situated on the forehead and directed upward. It was universally present in the oldest ostrachoderms. The third eye was smaller than the paired eyes and was also of general occurrence among ancient amphibians and reptiles. In living lizards the third eye is retained on the top of the head. It is buried under the skin having a miniature cornea, lens and retina that allow it to detect light (Romer and Parsons 1978).

Fig. 8.1 Eyes can aquire the most unexpected locations. Smooth hammerhead *Sphyrna lewini*. This shark species has eyes at the end of wing-like flaps

In other organisms eyes may also be located differently. The larval form of the widespread black dragonfish *Idiacanthus fasciola* has elongated stalks arising from the head, the eyes being situated at the extremities of these horn-like appendages. Among the sharks the location is still more extreme. The adult smooth hammerhead *Sphyrna zygaena* has eyes at the end of wing-like flaps that protrude from both sides of the head (Fig. 8.1).

Instead of two a fish may have four eyes. This is why *Anableps anableps* is called the "four eyes" fish. Its eyes have divided pupils and retinas enabling it to see above and below the water's surface simultaneously. Consequently it can keep a look-out for predators as it searches for food. A variant of this solution is found in the spookfish *Dolichopteryx binocularis*. As the name indicates its eyes have a most unusual anatomy being tubular in shape and pointing obliquely upwards like a pair of binoculars. This fish is considered to be the truly 4-eyed fish, because on the side of each tubular eye there is a second eye complete with retina and lens (Burnie 2004).

A phenomenon described since the early days of biology is the migration of eyes in flatfish during the transition from juvenile to adult. The hatching young of *Solea vulgaris* differs little from other fishes in the lateral position of its eyes, but at an early age the animal begins to lean to one side, at the same time that one eye migrates across the top of the head to take up a position beside the other eye. As the flatfish reaches the adult stage both eyes lie on the upper side of the body. The rotation may be to the left or to the right. In dextral fishes the left eye migrates, in the sinistral type it is the right eye that moves (Paxton and Eschmeyer 1998).

8.5 The Same Molecule (Retinoic Acid) Has a Primary Role in Both Eye and Body Development Disclosing the Fundamental Role of the Eye in Body Formation

Retinoic acid is a small lipophilic molecule derived from vitamin A (also called Retinol). This molecule is in many ways different from other diffusible cell-cell signalling factors that direct embryonic development. In contrast to protein factors (such as *fibroblast growth factor* and *hedgehog*) which bind to cell-surface receptors, retinoic acid enters the nucleus and binds directly to target genes (Duester 2008). During early development retinoic acid influences the patterning of embryonic tissues by organizing the trunk and at later stages contributes to the development of the eye.

In the mouse two distinct phases of retinoic acid signalling are required for eye development. In a first phase it leads to the formation of the neural retina and the optic cup. In a second phase retinoic acid travels outside the retina where it intervenes in the formation of another eye component—the cornea. Moreover, it was demonstrated that retinoic acid guides morphogenetic movements of neighboring cells. Experimentation involving transgenic mice has disclosed that the vertebrate body plan is mainly directed by an interaction between retinoic acid and *Hox* genes. The anterior regions of the body are generally more sensitive to retinoic acid indicating that a gradient exists along the anterior-posterior body axis (Marshall et al. 1996).

This molecular evidence obtained from mice strongly supports the data obtained from the molecular geometry of birds in which the eye was found to be a dominating and guiding center of body pattern spreading its influence to other organs of the bird's body (Lima-de-Faria 2012).

8.6 Eye Lenses Grow Like Tree Trunks by Adding New Layers of Cells

Most teleost fish, such as salmon, have spherical crystalline lenses. In Cichlid fish, of South and Central America freshwaters, the eye lens is composed of several hundred cell-layers, each layer consisting of thin elongated fibre cells. The younger, more peripheral fibre cells elongate as they mature. "The growth of the lens can be compared to that of a tree trunk: the oldest part of the lens is situated in the center and newer layers are constantly added in the periphery" (Schartau 2010) (Fig. 8.2).

The lens continues to grow throughout the entire life of the animal. Significant is that the primary lens fibre cells lose their organelles during the first stages of development. Most lens cells contain neither mitochondria nor other organelles. Despite this situation there is a constant input of energy based on glycolysis and an electric current which circulates in the lens (Schartau 2010). The lack of cell organelles is a condition necessary to ensure transparency, otherwise the light could not traverse the lens. The same is the case in humans where the cells of the eye lens lack nuclei and other cell organelles (Koretz and Handelman 1988).

8.7 The Circularity of the Vertebrate Eye

Over 8,000 species of birds have been described. An extensive study of these species led to recognize that the eye of birds has a particular circularity: (1) The skull contains a scleral ring consisting of small plates distributed as a perfect circle in which the eye is encased. (2) Bird specialists have described in detail two concentric rings which in several species surround the iris. They are called the eye-ring and the orbital ring. (3) This circularity extends to the whole head starting with an increase in the diameter of concentric rings formed by the regular distribution of pigment in the feathers. These rings extend to the head's periphery. (4) Remarkably this circularity of pigment distribution extends to the neck, breast, tail and wings in the form of larger circles (Lima-de-Faria 2012) (Fig. 9.2). A search was thus made to find out to what extent the circularity originating in the bird's eye was also a feature of other vertebrates.

It turned out that the eye has been incased in a circular skeleton since the appearance of the fish. The scleral ring that is preserved in the bird's skull also occurred in fossil fishes (*Rhabdomerma*), fossil amphibians (*Phlegethontia*) and fossil reptiles (*Sphenodon*). Mammals such as the fossil camel *Poebrotherium* and early humans (*Homo habilis*) had a skull with distinct circular eye cavities.

In present day living mammals, the circularity of the eye is accompanied by external concentric rings in the: Western tarsier, Spectral tarsier, the rodent speke's gundi and Guenther's Dik–dik.

The reptiles are in many ways as spectacular as the birds in their display of geometric patterns. The eyes of chameleons have a unique radial symmetry. The spheroid eye of the chameleon Knysna dwarf shows a single stripe that crosses the eye lid at its center and that has a horizontal position. Two other species (Malagasy and Johnston's chameleon) display 12 dark stripes which irradiate from a well-delimited center. In the Flap-necked chameleons eight distinct stripes follow the same radial pattern.

The amphibians are more modest in their geometry. The pupil, which is the aperture of the iris, has a slit shape that may occupy a vertical or a horizontal position in the eye of frogs. The Cape ghost frog has four dark stripes building angles of 90° that meet at the center of the eye.

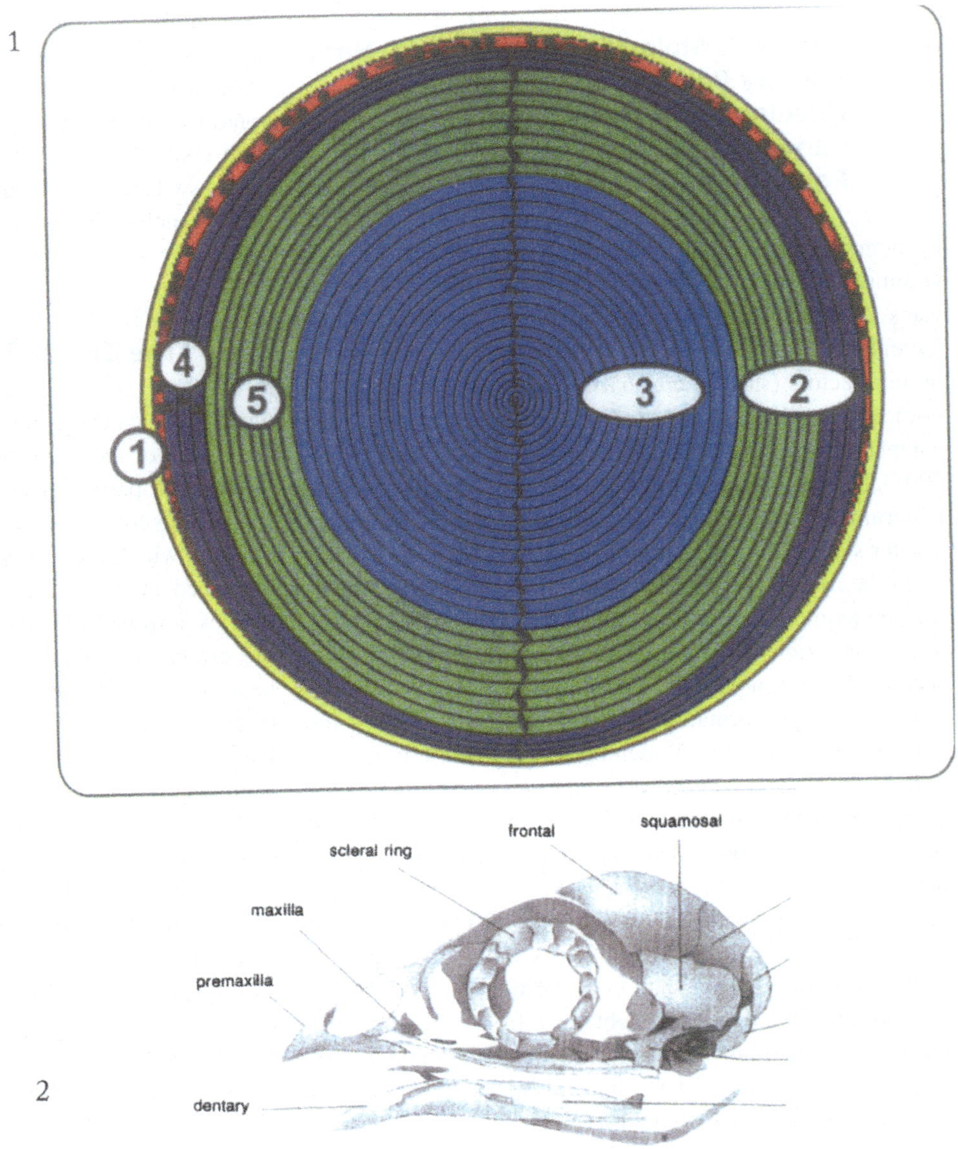

Fig. 8.2 **(1) Most teleosts have spherical crystalline lenses.** An eye lens belonging to a Cichlid fish. An exterior fibrous capsule and an epithelium layer enclose these lenses (1). The lens is composed of several hundred cell-layers, each layer consisting of *thin elongated* fibre cells. Lens cortex (2), lens core (3), differentiating fibre cells (4), mature fibre cells (5). The *circularity* of the fibre cells starts from a central point. The lens is growing continuously during the life of the animal, the oldest part is situated in the centre and newer layers are constantly added to the periphery. **(2)** ***The eye has been encased in a circular skeleton since the appearance of the fishes.*** Skull of a living bird

Fish have both extra rings around the eyes like some mammals and they may have radial stripes arising from the center of the eye as in frogs and chameleons.

The unity of body pattern among vertebrates shows up in many different ways and at unexpected occasions during evolution.

8.8 Circularity Extending from the Eye to Other Parts of the Head in Mammals

The human eye region, as well as the human mouth, have the capacity to change their shape in ways that allow humans to express their feelings in a unique way. The mouth, by changing its adjacent musculature produces the smile, a unique feature among mammals. The eye region with its circularly distributed muscles permits the expression of emotions with remarkable significance. The chimpanzee, the saki and *Cercopithecus* monkeys, display areas around the eyes which are partly circular (Fig. 8.3).

8.9 Genetic Intervention Has Produced Flies with Extra Legs and Extra Wings

At the same time that the eye remained a morphologically and functionally well-defined organ, it has evolved taking many different shapes, has occurred in various body locations, and has varied in number.

The genetic basis of this constancy allied to this variation, has recently been elucidated by gene manipulation in the fly *Drosophila*.

Already in the 1930s mutants had been described in which flies had legs located on their heads (Fig. 8.4).

Fig. 8.3 Geometric figures encircling the eyes of birds and primates. (**1**) Spotted Wood-owl *Strix seloputo* with two distinct *circles* surrounding the eyes. (**2**) Male white-faced saki *Pithecia pithecia* with its two *white half-circles* covering the frontal part of the head. (**3**) and (**4**) Frontal views of *Cercopithecus* monkeys (*C. cephus* and *C. ascanius schmidti*) displaying *blue* areas around eyes which are partly *circular*

This finding remained mainly a curiosity, although at the time several geneticists pointed out its significance. They stressed that this result indicated that much remained to be learned about development and evolution.

One had to wait until the 1970s and 1980s for a large number of laboratories in Europe and the USA to tackle this problem in depth.

The results were astonishing (Fig. 8.4).

A normal fly has six legs, which is the regular number in all described insect species (over one million). Yet flies with eight legs were produced in the laboratory. This is the same number of legs that is typical of spiders which the classification of invertebrates locates in another class. The emergence of eight legs was obtained using a fly genotype with a *bithoraxoid/Df* gene combination, where *Df* is a deficiency in the *bithoraxoid* gene.

In the antenna to leg transformations the following genetic route was followed. Transgenic flies were obtained by fusion of the heat-shock promoter to the gene *Antennapedia*. Initially the transgenic flies had normal antennae but, by refining the experiments, large legs with claws appeared on the head (Fig. 8.4), Gehring (1998).

The number of wings was also changed from 2 to 4. In flies found in nature the third thoracic segment forms only small balancers instead of wings. As a result normal flies have only two wings. By inactivating the *ultrabithorax* genes, it became possible to convert the balancers into a second pair of wings producing a fly with four well-shaped and large wings (Bender et al. 1983; Lawrence 1992). Butterflies are insects whose body consists regularly of four wings. Hence, one was recreating in the laboratory a situation which occurred in nature in other insect species. Actually evolution was being repeated at will by experimental means.

This gene manipulation was extended to vertebrates with similar results. Following the same technology, birds with four legs and four wings were obtained experimentally (Rodriguez-Esteban et al. 1999; Cohn et al. 1997; Isaac et al. 1998) (see Fig. 5.3 in Lima-de-Faria 2012).

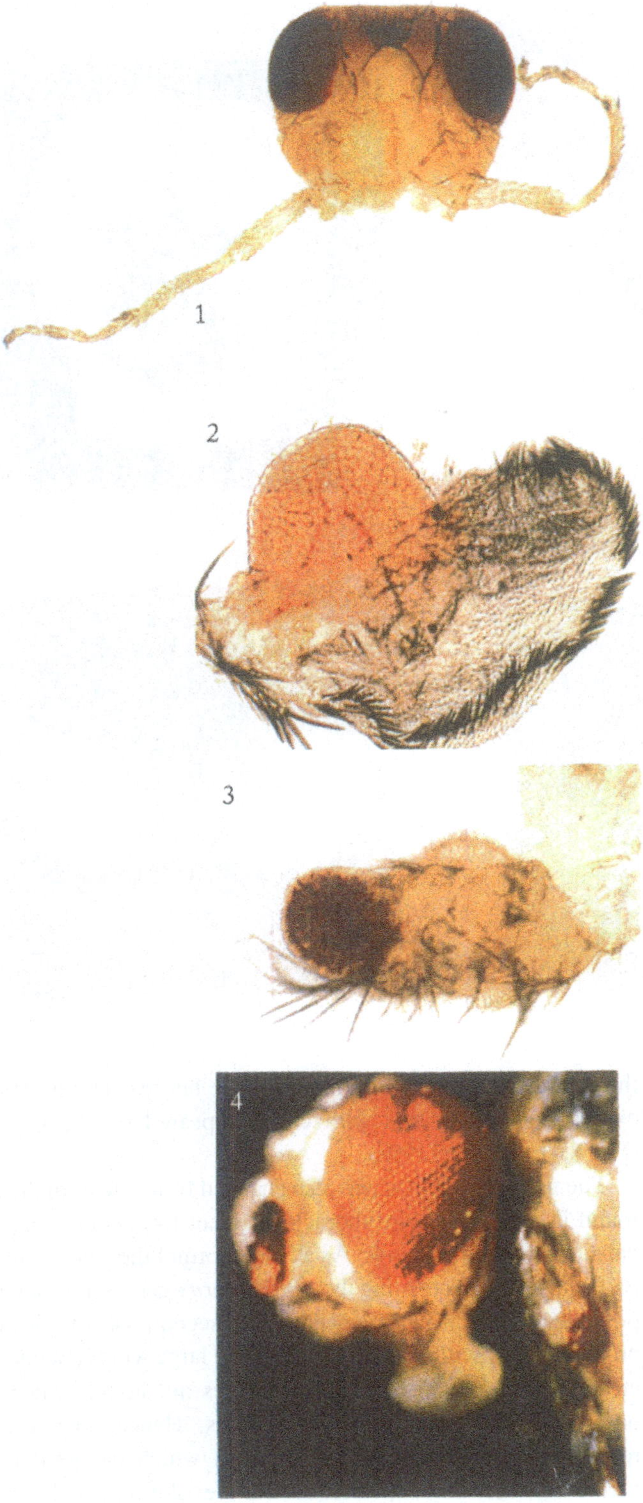

Fig. 8.4 **Genetic manipulation in the fly *Drosophila* has led to the production of extra organs in other body parts revealing unity and simplicity of organization**. (**1**) Fly with two legs on head. In the homeotic mutant *Antennapedia* the antennae are transformed into middle legs. (**2**) Large eye on wing. Induction of additional eyes on the wings of a transgenic fly by expression of the normal *eyeless* gene. (**3**) Eye on antenna. Antenna in which most of the third antennal segment is replaced by eye structures. (**4**) Large normal eye plus two small eyes on antenna and foreleg. Transgenic fly with additional eyes on the antenna and the foreleg by expression of the normal *eyeless* gene

8.10 Unexpectedly the Eye Turned Out to Be Able to Occupy the Most Extraordinary Positions Revealing Plasticity and Unity of Body Organization

Outside these gene clusters other homeobox genes have been discovered, the most interesting being *eyeless*, which corresponds to the mutation *Small eye* in mice and to *Aniridia* in humans (Quiring et al. 1994). Following a series of experiments, transgenic flies were obtained in which eyes were formed on the wings, legs and antennae (Halder et al. 1995). The eyes are real eyes because they are identical in structure and contain functional photoreceptors. This means that a single master control gene can start the cascade of eye development which involves not less than 2,500 other genes needed for eye formation.

The fact that eyes can be formed in most areas of the body, and are true morphological and functional structures, and that little wings can form in the eye, implies a remarkable unity of body construction. Hence, the eye cannot be separated from the wing or from the leg (Gehring 1998).

Corroborating these results was the finding that mouse genes produce eyes in flies. Mammalian *Hox* genes have turned out to be functional homologues of the homeotic *Drosophila* genes, as demonstrated by an experiment in which *Hox* genes of the mouse were introduced into flies. The mouse vector induced antenna to leg transformation in adult flies. Moreover, the mouse *Pax-6* (*Small eye*) gene was able to induce eyes in the fly (Halder et al. 1995; Callaerts et al. 1997). There is an extensive nucleotide sequence homology between the fly *eyeless*, the mouse *Small eye* and the human *Aniridia* genes.

These findings give more weight to the primary role of the eye in shaping body structure and determining evolution.

Sources of Figures

Fig. 8.1 Smith, M.M. and Heemstra, P.C. (Editors) 1986. Smiths' Sea Fishes. Springer, Berlin (Fig. Plate 2, Nr. 13.1).

Fig. 8.2 (**1**) Schartau, J.M. 2010. Morphological and Physiological Adaptations in Cichlid Crystalline Lenses. Doctoral Thesis, Lund University, Department of Biology, Sweden (Fig. 5, page 17). (**2**) del Hoyo, J. et al. (Editors) 1992. Handbook of the Birds of the World, Vol. 1. Lynx Edicions, Barcelona, Spain (Vol. 1, Fig. page 43).

Fig. 8.3 (**1**) del Hoyo, J. et al. (Editors) 1999. Handbook of the Birds of the World, Vol. 5. Lynx Edicions, Barcelona, Spain (Fig. Vol. 5, Plate 12, Nr. 94). (**2**) Napier, J.R. and Napier, P.H. 1985. The Natural History of the Primates. British Museum, London, UK (Photo page 120). (**3**) + (**4**) Napier, J.R. and Napier, P.H. 1985. The Natural History of the Primates. British Museum, London, UK (Figs. page 141).

Fig. 8.4 (**1**) Gehring, W.J. 1998. Master Control Genes in Development and Evolution: The Homeobox Story. Yale University Press, London, UK (Plate 1). (**2**) Gehring, W.J. 1998. Master Control Genes in Development and Evolution: The Homeobox Story. Yale University Press, London, UK (Plate 7). (**3**) Halder, G. et al. 1995. Induction of ectopic eyes by targeted expression of the *eyeless* gene in *Drosophila.* Science 267: 1788–1792 (Fig. 2, page 1789). (**4**) Gehring, W.J. 1998. Master Control Genes in Development and Evolution: The Homeobox Story. Yale University Press, London, UK (Plate 7).

The Circularity of the Vertebrate Body 9

9.1 Circularity in Fossil and Living Mammals

The general image of a mammal is that of an animal standing on its four legs having a narrow trunk connecting the head with the tail. The main body configuration conveys no sense of circularity.

However, not all mammals that populated the earth in earlier periods and those which exist today convey this image. During the Pleistocene (1 million years ago) a most unique animal lived among the diversified fauna of South America. The largest specimens of the giant *Glyptodon* were up to 2.5 m long not including the tail. The main part of its body was formed by a nearly spherical and inflexible shell, consisting of hinged plates of bone (Halstead 1978, Fig. 9.1).

The *Glyptodon* belonged to the armadillo line. Thus, it is not surprising to find in present day South America the living Patagonian armadillo *Zaedyus pichiy* (also called Pichi). The main part of the body is a near circular armor consisting of plates of bone with horny covering. These are grouped in eight hinged bands. The head and the rear are covered by plates as in the fossil *Glyptodon* (Burnie 2004).

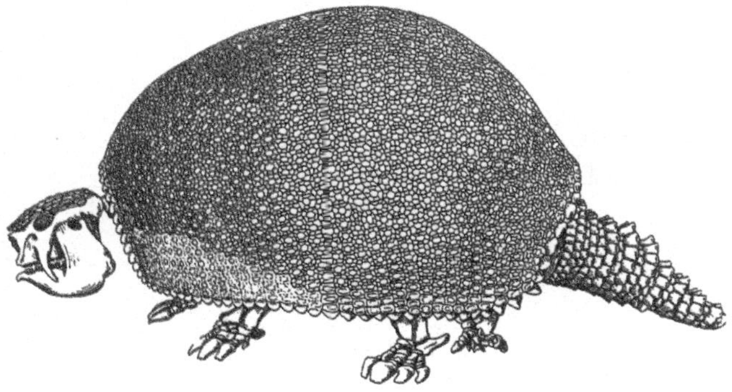

Fig. 9.1 The *circularity* of the body in fossil mammals. Giant *Glyptodon* from the Pleistocene of South America. It belonged to the armadillo line. The largest animals were up to 2.5 m long not including the tail. The inflexible shell consists of hinged plates of bone

9.2 Circularity in Birds Takes Another Form Due to the Presence of Feathers

The situation in birds is different from that in mammals, reptiles, amphibians and fishes. In these last groups the shape of the body may actually be quite close to that of an ellipse or a circle, but in birds it is not the body itself, but the pattern that the pigments build on the feathers, that exposes a distinct circularity. This regular feather coloration can be seen extending from the eye to the head, the wings and to the whole body (Fig. 9.2).

The bird skeleton has a compressed form. When the feathers are removed, the central part of the body is round as compared with the oblong trunk of most mammals.

9.3 Turtles and Rays Tend to Be Highly Circular and Geometric

Reptiles exude geometry. The North American spiny softshell turtle *Apalone spinifera* is characterized by three geometric traits: (1) The shell is mainly elliptical or oval. (2) A brown line is concentric with the periphery of the shell. (3) The whole carapace is covered with brown circles of various sizes, their distribution tending to follow a gradient. The largest are located at the center of the shell, as their size decreases successively the smallest lie on the periphery except for a few of minute size which appear throughout the carapace.

A. Lima-de-Faria, *Molecular Origins of Brain and Body Geometry*,
DOI 10.1007/978-3-319-06056-9_9, © Springer International Publishing Switzerland 2014

Fig. 9.2 ***Circularity* in birds.** (**1**) Asian Barred Owlet *Glaucidium cuculoides*. A most striking example of total body *circularity*. *Concentric circles* start surrounding the eyes, extend to the whole of the head, expand into the breast, cover the wings and the tail. In all cases the stripes tend to be parallel and spaced at regular intervals. (**2**) Sjöstedt's Owlet *Glaucidium sjostedti*. Another closely related species in which the circularity extends to the whole body. In this case, the *concentric rings* that start around the eyes can be seen to extend into the back of the head. Parallel stripes, distributed at regular intervals, cover the feet, showing that the *body circularity* extends from one extreme to the other of the avian body. (**3**) Great Hornbill, *Buceros bicornis* showing the *circularity* of the wings and the tail. (**4**) Guianan Cock-of-the-rock *Rupicola rupicola*. The *circularity* of the crown is accompanied by a *concentric dark semicircle*. (**5**) Double-barred Finch *Taeniopygia bichenovii*. The *black circle* around the eye is accompanied by another one surrounding the breast. (**6**) Lady Amherst's Pheasant. The *circularity*, that starts at the eye, extends to most parts of the body. Parallel transversal stripes, located at regular intervals, are seen on the wings and along the whole tail

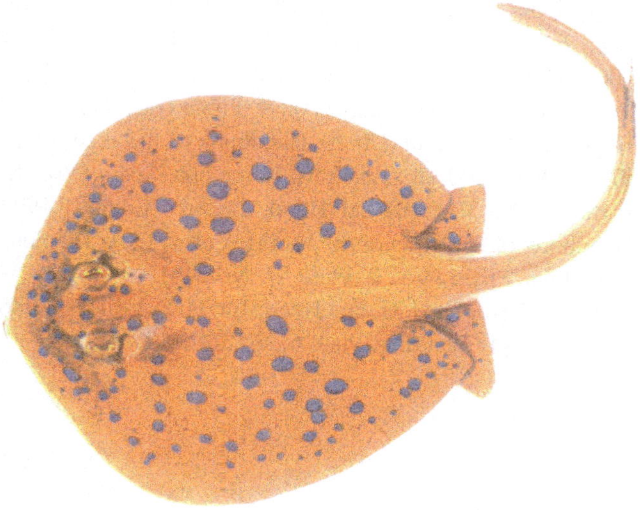

Fig. 9.3 The body of fishes can also be *circular*. The ray *Taeniura lymma* with a body shape between an *ellipse* and a *circle*. The pelvic fins are quite small

The Cogwheel turtle *Heosemys spinosa*, as the name indicates, has the shape of a wheel but is also close to an elliptical form. The hexagonal and pentagonal plates of the carapace add to its geometry as well as the yellow band that functions as a diameter separating the two halves of the carapace.

The line, that builds the diameter of a circle, occurs in the circular body of the Mexican burrowing frog *Rhinophrynus dorsalis*. In this amphibian the line is red stretching itself from the head to the rear being well delimited against a black body background.

Most fish have a fusiform body, but there are species which could hardly be more circular. The ocellated freshwater stingray *Potamotrygon motoro* has a circularity derived from five main traits: (1) The body's shape is between an ellipse and a circle. (2) Three rings of small circles, having a constant dimension, are distributed concentrically along the body's periphery. (3) The central region of the body is covered by larger black and yellow circles which are regularly distributed. (4) These large circles extend to the whole tail where they get a central black dot. (5) The circularity is so extreme that the pelvic fins are hardly seen.

Another species of ray *Taeniura lymma* has also a body, covered with dark spots, which approaches circularity. As in the previous case the larger dots occupy the central part of the body but their distribution is less regular and the small pelvic fins protrude (Fig. 9.3).

Another ray has a name that betrays its circularity. The round stingray *Urolophus halleri* has light circular patches that occupy the body's central region whereas a few smaller ones are in the periphery. Besides, minor white dots are distributed over the whole fish.

Taken together, these findings show that among vertebrates a body shape, that is mainly circular, occurs in given species of mammals, reptiles, amphibians and fish. It is significant that in the birds this circularity was transferred to the feathers where it is most distinct.

Sources of Figures

Fig. 9.1 Boule, M. and Piveteau, J. 1935. Les Fossiles (Fig. 1204, page 744).

Fig. 9.2 (**1**) del Hoyo, J. et al. (Editors) 1999. Handbook of the Birds of the World, Vol. 5. Lynx Edicions, Barcelona, Spain (Vol. 5, Plate 16, Nr. 143). (**2**) del Hoyo, J. et al. (Editors) 1999. Handbook of the Birds of the World, Vol. 5. Lynx Edicions, Barcelona, Spain (Vol. 5, Plate 16, Nr. 142). (**3**) del Hoyo, J. et al. (Editors) 2001. Handbook of the Birds of the World, Vol. 6. Lynx Edicions, Barcelona, Spain (Vol. 6, Plate 39, Nr. 28). (**4**) del Hoyo, J. et al. (Editors) 2004. Handbook of the Birds of the World, Vol. 9. Lynx Edicions, Barcelona, Spain (Vol. 9, Photo page 42). (**5**) del Hoyo, J. et al. (Editors) 2010. Handbook of the Birds of the World, Vol. 15. Lynx Edicions, Barcelona, Spain (Vol. 15, Plate 25, Nr. 94). (**6**) del Hoyo, J.et al. (Editors) 1994. Handbook of the Birds of the World, Vol. 2. Lynx Edicions, Barcelona, Spain (Vol. 2, Plate 57, Nr. 143).

Fig. 9.3 Smith, M.M. and Heemstra, P.C. (Editors) 1986. Smiths' Sea Fishes (Fig. Plate 6, Nr. 30.13).

10 Concentric Circles Are a General Feature of Vertebrates

10.1 Animals with Drab Colors Dominate in the Northern Hemisphere

The vertebrates that we are most familiar with, i.e. those that inhabit the Northern latitudes are usually characterized by uniform drab colors. The bear and the mouse are usually grey, eagles and sparrows are mostly brown, a snake's body is dominated by black and grey, toads are usually dark, and fishes like the cod and herring have silvery tones.

As one enters the tropical regions the situation changes dramatically. Striking colors and exquisite markings not only appear in birds, where they are most evident, but are also present in fish, amphibians, reptiles and mammals.

It has been demonstrated experimentally that temperature affects body coloration (Chap. 22).

10.2 Large Concentric Circles Have Particular Locations Along the Body of Fishes and Turtles and Occur As Well in Other Vertebrates

It is not by accident that a turtle has been named Peacock soft-shelled turtle. The upper part of its carapace displays four large rings. Each ring consists actually of five concentric circles of alternating light and dark color. The tail of the peacock bird is covered by over a dozen rings built similarly by concentric circles of alternating dark and light colors (Fig. 10.1).

An additional detail makes the shell of this turtle exceptional. The four concentric circles are not disposed erratically on the carapace but are located symmetrically, their centers being at the vertices of a square (Fig. 10.1).

Even more distinct circles are present in the Clown knifefish. Twelve black solid circles, each surrounded by a well-delimited light circle, are found along the whole body with the exception of the head. Their centers are distributed along an undulating line that finishes just before the caudal fin. Another fish, mandarin, derives its name from the beautiful colored dress of Chinese mandarins. Its caudal and dorsal fins are covered by yellow, black and blue concentric circles that rival those of the peacock.

There are also large conspicuous circles which occupy particular body locations. Some are near the head others are at the end of the body. A single solid black circle, accompanied by a concentric large white circle, is present near the head in Dory fish and Port Jackson shark. A solid grey circle delimited by a black concentric circle appears also near the head of the Lizard male banded anole.

On the rear, black circles, encircled by white or red rings, occur in False-eyed frog and Velvet cichlid (Fig. 10.2).

The location, of these circles in the frontal and rear areas of the animal body, is not accidental. This is disclosed by the evidence gathered on reptiles and mammals described in the following chapter.

A. Lima-de-Faria, *Molecular Origins of Brain and Body Geometry*,
DOI 10.1007/978-3-319-06056-9_10, © Springer International Publishing Switzerland 2014

Fig. 10.1 ***Concentric circles* at specific body parts.** (**1**) The fish *Callionymus maculatus* with its dorsal fins covered by *concentric circles*. (**2**) Peacock soft-shelled turtle *Trionyx hurum*. It has four symmetrically located *large rings* consisting of 3 *black* and 2 *white* circles which are all concentric. The similarity of the arrangement of these *circles* with those of the peacock is so striking that it became part of its name. (**3**) Green peafowl *Pavo muticus* (peacock). The numerous rings on the long feathers of the tail are formed by four *concentric circles*: *black, blue, orange* and *black*

Fig. 10.2 ***Solid circles* located close to the rear.** (**1**) False-eyed frog *Eupemphix nattereri*. In this amphibian the *solid black rings*, with a *concentric white circle*, are located on the rear. (**2**) The fish *Chaetodon lunula*. The two rings appear like moons as the latin name indicates

Sources of Figures

Fig. 10.1 (**1**) Nationalnyckeln till Sveriges Flora och Fauna. Ryggsträngsdjur. 2012. Artdatabanken. Sveriges Lantbruksuniversitet (Fig. page 389). (**2**) Pritchard, P.C.H. 1979. Encyclopedia of Turtles. T.F.H. Publications, Neptune, New Jersey, U.S.A. (Fig. page 393). (**3**) del Hoyo, J. et al. (Editors) 1994. Handbook of the Birds of the World, Vol. 2. Lynx Edicions, Barcelona, Spain (Vol. 2, Plate 59, Nr. 154).

Fig. 10.2 (**1**) WIKIPEDIA.org/WIKI. Author Felipe Gomes. (**2**) Smith, M.M. and Heemstra, P.C. 1986. Smiths' Sea Fishes. Springer, Berlin [Fig. Plate 75, Nr. 205.9(A)].

11 Two Poles of Circularity: Head–Neck and Rump

11.1 In Reptiles Neck and Rump Have a Particular Circularity Not Connected with the Skeleton

Among vertebrates the common flying dragon *Draco volans*, of South East Asia, has a unique circular pattern that starts around the neck. This reptile does not fly (as the name states) but only glides. During gliding this species opens up its elongated ribs which are connected by a membrane that is a half circle. Five distinct black bands build semicircles starting at the neck. They are concentric and increase in diameter with increasing distance from the neck. They extend as far as the rump where they terminate.

One would expect that this band coloration would follow the location of the long ribs. But this is not the case. On the contrary the ribs, which also tend to have a semicircular distribution have an opposite orientation. The minor semicircle built by the smallest ribs starts at the rump and as the ribs increase in size they reach the neck (Fig. 11.1).

Hence, the coloration pattern not only is independent of the skeleton but has a distinct opposite distribution. However, both follow their own circularity, one oriented to the neck the other to the rump.

11.2 Head–Neck and Rump Display Identical Circular Patterns in Mammals

Careful analysis discloses that among mammals there are species with a distinct neck circular pattern. These are found in most diverse families: Tiger, drill, Iberian lynx, Grevy's zebra, Northern bottlenose whale, Collared peccary and Asian wild buffalo. The rings around the neck of the drill, the zebra and the whale are most distinct and tend to be accompanied by others which are concentric (zebra and tiger) (Fig. 11.2).

What is surprising is that the rump displays also a clear-cut circularity in mammal species like: Przewalski's gazelle, Ellipsen waterbuck, mandrill and drill. In the waterbuck two white semicircles on the rump are accompanied by a white semicircle surrounding the neck. This occurs both in females and males. Again the rump and head–neck have similar patterns as is the case in the drill (Fig. 11.3).

The unity of pattern in vertebrates surfaces again. Some birds have beautifully colored concentric rings around their necks, whereas others have white circular marks on their rump (Figs. 11.2 and 11.3). The same is true of fish. The circularity of the rump is especially evident in some fish. White concentric circles of increasing diameter start at the rump of the Emperor angelfish *Pomacanthus imperator* (juvenile form) (Fig. 11.3).

Among the zebra species there is one which presents this polarity of pattern in an unexpected way. It shows parallel black and white stripes that encircle the neck. Moreover, the rump of the Grevy's zebra *Equus grevyi*, is different. Just where the tail starts, a series of small concentric black and white stripes are formed that, like in the angelfish, increase in diameter as they get further away from their initial centre at the base of the rump (Fig. 11.4). Like in the waterbuck the circularity of the rump is accompanied by an equally distinct circular pattern around the neck. This is best seen in Fig. 11.2 where the animal is photographed from a frontal position.

Anatomists know that the human body has a particular distribution of peripheric nerves just around the rump as depicted in anatomy treatises (Fig. 11.4). This similarity may not be fortuitous. The back of a woman's body is known to be an erogenous zone, a trait that may have an ancestral origin derived from the mating behavior common to most mammals.

Amphibians have been extensively used in experiments designed to study the distribution of the bioelectric field in animals. Adult newts were employed to measure the total

A. Lima-de-Faria, *Molecular Origins of Brain and Body Geometry*,
DOI 10.1007/978-3-319-06056-9_11, © Springer International Publishing Switzerland 2014

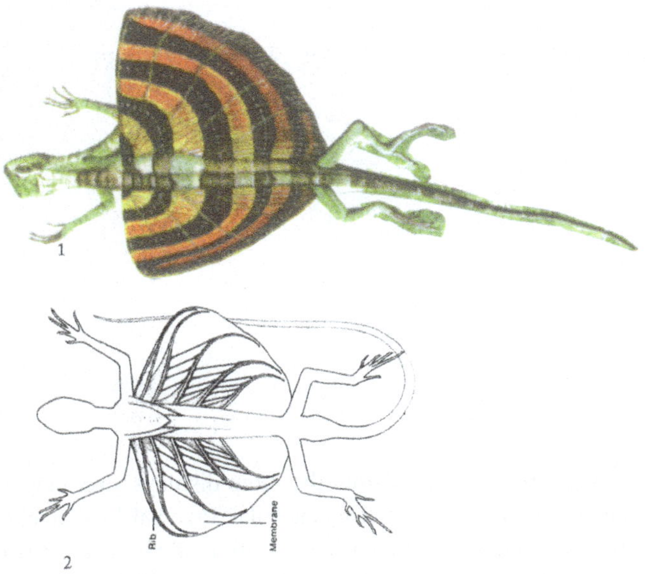

Fig. 11.1 ***Concentric semicircles*** **on the body of reptiles starting close to the neck region.** (**1**) Common flying dragon *Draco volans* of South East Asia. This reptile does not fly but glides. (**2**) Flying lizard of South East Asia showing the supporting ribs which are joined like those of an umbrella. When not required, the membranes can be folded down against the animal's sides

body potential by inserting recording electrodes at tip-of-nose and at tip-of-tail. Remarkable is that the electrical field distributed itself in the form of parallel lines in the head, legs, trunk and tail, but it formed distinct concentric circles on the neck and rump. As pointed out by Berrill and Karp (1976) these two centers are one adjacent to the brain and the other to the sexual organs. The directing capacity of the brain and the perpetuation of the species by reproduction are the two most vital features of the body.

11.3 The Two Poles of Circularity Seem to Have a Genetic Basis

Nothing is accidental in a vertebrate body. Every trait is the product of a gene-environment interaction that is directed by a main molecular cascade.

The presence of two circular poles close to the two extremities of the body seem so novel, and for this reason so odd, that at first sight they are expected to arise from nowhere.

But, as found previously, when molecular genetics advances, the source of every trait, may it be a part of the skeleton, a physiological process, or the geometry of coloration, turn out to have their origin in gene action.

It turns out that these two geometrical and physiological poles seem to be directed by the same genes as disclosed by studies on mice. The skeleton of vertebrates is under the control of the *Hox* genes which decide the segmentation of the embryo into the anterior, median and posterior sections of the vertebral column (Young and Deschamps 2009). Another group of genes (*Dix5* and *Dix6*), which are common to mammals and flies, were found to regulate the skeleton of mice. The simultaneous mutation of these two genes resulted in the vertebral column being formed, but the skull and the terminal region of the tail were absent. Thus, the two extremities of the vertebrate skeleton are under the control of the same genes (Kraus and Lufkin 2006), but more genetic evidence is necessary to clarify this relationship.

11.4 The Main Geometric Features of the Vertebrate Body Can Be Condensed in Graphic Form

The analysis of the molecular geometric pattern of birds led to synthesize this information on a circular diagram in which the eye occupies its center (Fig. 11.5). Obviously, such a figure represents a simplification but it has the advantage of concentrating the geometric figures, recognized in the different parts of the body, into a unitary structure, because it turned out that the formation of concentric circles started at the eye and expanded successively to all parts of the body including the bill and the legs. This unity of pattern organization was reinforced by genetic manipulation in which birds with four wings and four legs were obtained and by the production of flies (which have similar genes) with wings on eyes and of eyes on wings as well as the occurrence of legs on the head and of antennae on legs, a most impressive demonstration of the unity of body formation (Fig. 8.4).

The geometric patterns analysed in this work in the thousands of vertebrate species, that have been described in detail, reveal the same type of geometric figures found previously in birds.

A novel feature emerged when considering all the vertebrates, and specially the mammals. The eye continues to be the main center of the origin of the pattern, but is now counterbalanced by a second pole of circularity arising at the rear with its center at the site where the tail leaves the trunk (Fig. 11.4).

The body of most vertebrates is oblong but may be even circular or ellipsoidal. In the construction of the graphic representation the intermediate ellipsoidal form of the trunk was chosen as being representative. In the head the concentric circles are drawn emanating not from the two eyes present in vertebrates but from a single eye with a parietal position (also called third eye). This eye which was located at the top of the head in many vertebrate fossil species still has the same head location today in living lizards (and is important in solar orientation). It may be recalled that the eye, in living flatfish, can move to the top of the head during development.

Fig. 11.2 ***Circular figures*** **around the head–neck region.** (**1**) Tiger *Panthera tigris*. (**2**) Grevy's Zebra *Equus grevyi*. (**3**) Double-banded Courser, *Smutsornis africanus*. (**4**) Three-banded Plover *Charadrius tricollaris*

Two other novel traits are the dark thick band that runs along the vertebral column, from the head to the rear, dividing the main part of the vertebrate body into a left and a right side (Fig. 6.1). The second trait, of no less significance, is a vertical band that divides the body into an anterior and a posterior part, and is located in the middle of the trunk (Fig. 7.1).

The four legs characteristic of the mammals, reptiles, and amphibians are depicted (they tend to be narrower near the feet). In birds there are two legs and two wings and in fish these are substituted by fins of different numbers whereas snakes have no legs. Yet the unity of the vertebrate skeleton remains evident from fish to humans (Fig. 3.5).

No imagination was needed to create the graphic representation of the vertebrates. Like with the birds, all traits included were those that were found to be highly explicit and to be the most representative geometric figures that appeared where least expected (Fig. 11.6).

Fig. 11.3 **Rings (*mainly white*) on the rumps of mammals, birds and fishes.** (**1**) and (**2**) Female and male of Ellipsen Waterbuck *Kobus ellipsiprymnus*. In the waterbuck there is also a *ring* around the neck in both sexes. (**3**) Przewalski's Gazelle *Procapra przewalskii*. (**4**) Guadalupe Storm-petrel *Oceanodroma macrodactyla* with *white semi-circle* on rump. (**5**) Emperor angelfish *Pomacanthus imperator* (juvenile form) having a most conspicuous *circular rump*

11.5 The Construction of the Geometric Diagram Follows Solely the Ordered Pattern Found in Vertebrates

The same sets of genes that are present in primitive vertebrates can be recognized in human chromosomes (Fig. 3.1). Hence, it is not surprising that this permanence of organization is evident at the geometric level.

The numbers 1–12 in Fig. 11.6 refer to the following features:

1 **Bill**. Examples of structures protruding from the head, in a position opposing the tail, are the bills of birds, the trunks of elephants and the bill-like jaws of many fish species. The bills of birds are covered by as many as seven stripes in several species of Hornbills. These stripes are parallel to each other and are oriented in a transversal direction. The bills of other species show also parallel bands that follow the same orientation. Mammals have also bills as is the case with the Duck-billed platypus (Fig. 11.7).

2 and 3 **Forelegs** and 4 and 5 **hindlegs**. The two types of legs tend to have the same pattern of longitudinal stripes,

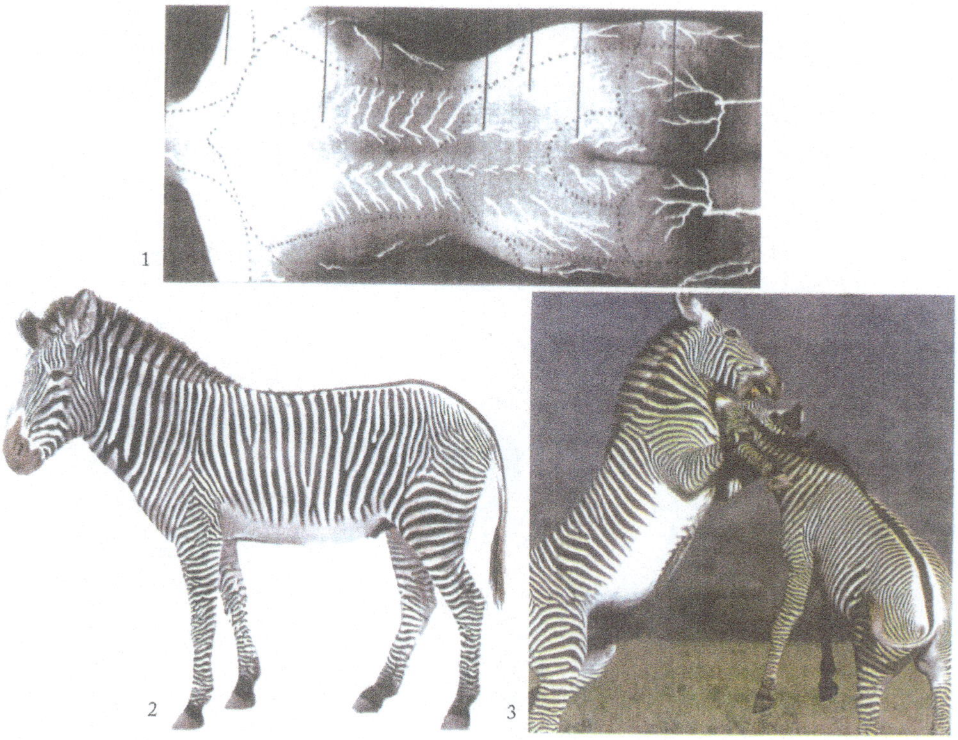

Fig. 11.4 ***Circles* on the rump of humans and zebras.** (**1**) Distribution of peripheric nerves in a woman at the central part of the rump giving evidence of the circularity of the sacral area. (**2**) Grevy's Zebra *Equus grevyi* with its characteristic *concentric rings* that start on the rump at the beginning of the tail. (**3**) The same zebra species, seen from the rear, showing better the *circular area* and the *long black stripe* along the dorsal rim

which are parallel, and most distinct in zebras (Fig. 11.4), but also in chameleons and the ringtailed gecko.

6 **Tail**. Parallel and transversal broad stripes covering the tail is an impressive sight in lemurs, genets and other mammals (Fig. 6.4) but it is equally evident in reptiles such as the Blue-tongued skink.

7 **Eye, primary pole**. In the diagram, instead of two lateral eyes a single eye is represented. As stated by Romer and Parsons (1978): "Ancestral vertebrates had a third eye, medially situated on the forehead and directed upward. This eye was generally present in the placoderms and present in Devonian bony fishes of all major groups. Further, it was universal in all the older land vertebrates—ancient amphibians and Paleozoic reptiles of all sorts".

8 **Rump, secondary pole**. The circularity of the rear or rump is most distinct in Grevy's Zebra (Fig. 11.4) but is equally impressive in the Ellipsen Waterbuck (Fig. 11.3).

9 **Heavy spine band**. A horizontal broad band runs along the vertebral column from head to tail (sometimes starting at the nose and in other cases prolonged into the tail). This band creates a sharp division between the left and right sides of the body. It is unmistakable from mammals to fishes (Figs. 6.1 and 11.8).

10 **Line dividing anterior from posterior region**. This vertical line appears in the middle of the body separating, in a sharp way, the anterior from the posterior regions of the trunk. It is distinct in several species of zebra, especially in *Equus burchelli* (Fig. 7.1) and fishes (Fig. 11.8).

11 **Anterior region of body**. This is covered by: circular lines (Figs. 9.2 and 11.8), vertical lines (Fig. 6.3) and horizontal ones (Figs. 6.2 and 11.7).

12 **Posterior region of body**. It is also covered by similarly oriented stripes (Figs. 9.2, 11.7 and 11.8).

It turns out, that all these traits are most evident in fish which are the oldest vertebrates.

11.6 The Third Eye Located on the Top of the Head in Reptiles "Represents Evolution's Earlier Approach to Photoreception"

Two surviving species of the ancient order Rhynchocephalia (*Sphenodon punctatus* and *Sphenodon guntheri*) have in addition to large lateral eyes a third eye located on the top of the head. They are living fossils, which at present inhabit New Zealand. Their heyday was over 200 million years ago

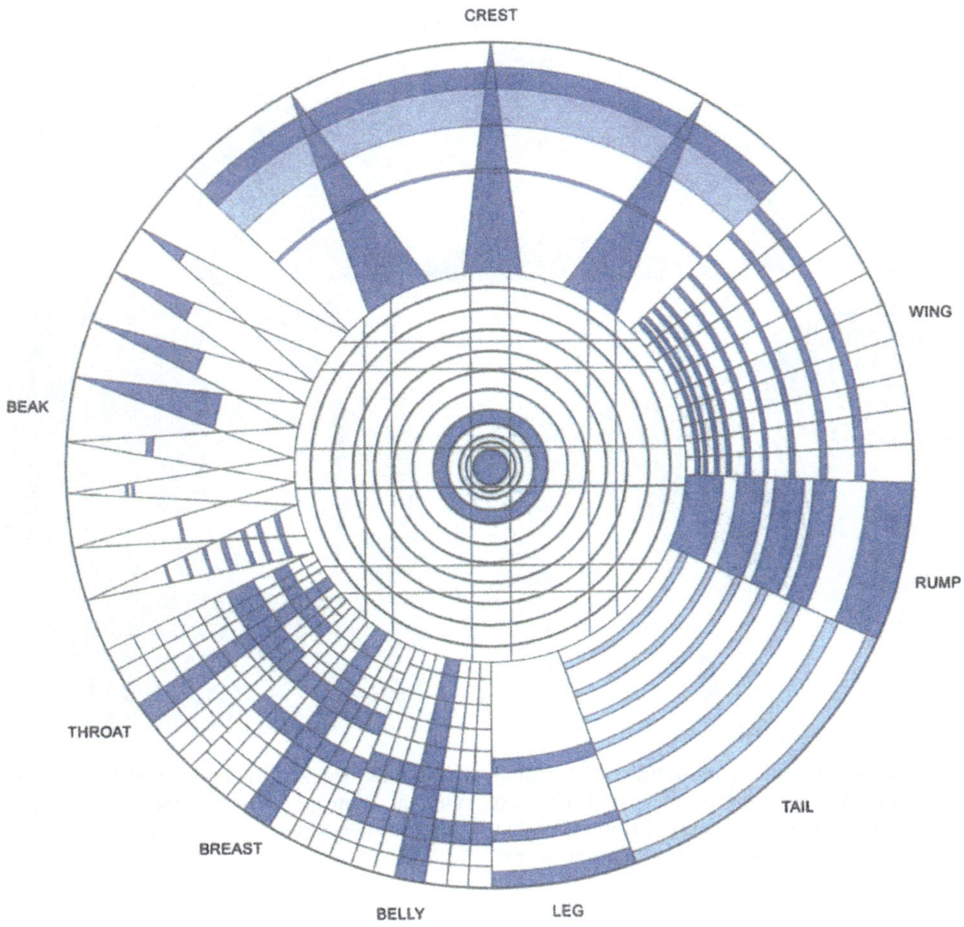

Fig. 11.5 Disk synthesizing the information on the geometry of body pattern collected from the analysis of over 8,000 species of birds. The eye occupies the central section of the disk, since it turned out to be the focal point from which depart: *concentric rings, semicircles, parallel lines*, and other *geometric figures*. This geometry is displayed by the whole avian body extending from the head's crest to the tail and the legs. In the disk the distribution of the bird's structural regions follows their natural position in the body which is reduced to a *circle*. The features depicted in the different structural sections of the disk are those that were found to be most evident in the bird species analysed. The genetic information available discloses that the pattern can be correlated with molecular cascades which originate in specific DNA sequences. These are responsible for the emergence and shaping of the body structures. Moreover, feather pigments, which are the main components of the concretization of the geometry, due to the figures that they form, turn out to be guided by proteins during their ordered migration. Additionally, genetic manipulation has led to the demonstration that eye genes are keys to the understanding of the unity of body organization. The geometric disk represents a simplification of the biological reality, just as every figure of Euclidean geometry is a simplification of the physico–chemical world. The simplicity of the disk may seem provocative but any basic phenomenon must be simple otherwise it would have never worked. All rights reserved covering this original figure. It can only be reproduced with permission of the author or his heirs, being accompanied by: (**1**) The complete text of the figure caption. (**2**) A citation as follows: Lima-de-Faria, A. 2012, "Molecular Geometry of Body Pattern in Birds", Springer, Berlin, New York, Fig. 8.36, page 112. (**3**) Consent will only be given for reproduction of the entire unmodified image

but they look like most other reptiles (Halliday and Adler 2004).

This third eye was lost in later orders such as crocodiles, birds and mammals, but remnants of this organ can be found in most of these groups. It "represents evolution's earlier approach to photoreception" (Schwab and O'Connor 2005). The third eye is a complex structure consisting of: a cornea, a lens (that is surprisingly similar to that of lateral eyes), vitreous cavity, pigment epithelial layer, ciliated photoreceptors and ganglion cell layers.

11.7 This Eye Functions As a Sky Polarization Compass that Senses Ultraviolet Light

The function of the third eye has been recently investigated in detail. It functions as a sky polarization compass in *Podarcis sicula*, *Sceloporus jarrovi*, *Tiliqua rugosa* and *Uma notata*. Following a series of experiments the lizards were trained to swim from the center of a water maze onto a hidden platform (the goal) positioned at the periphery of the maze in a single

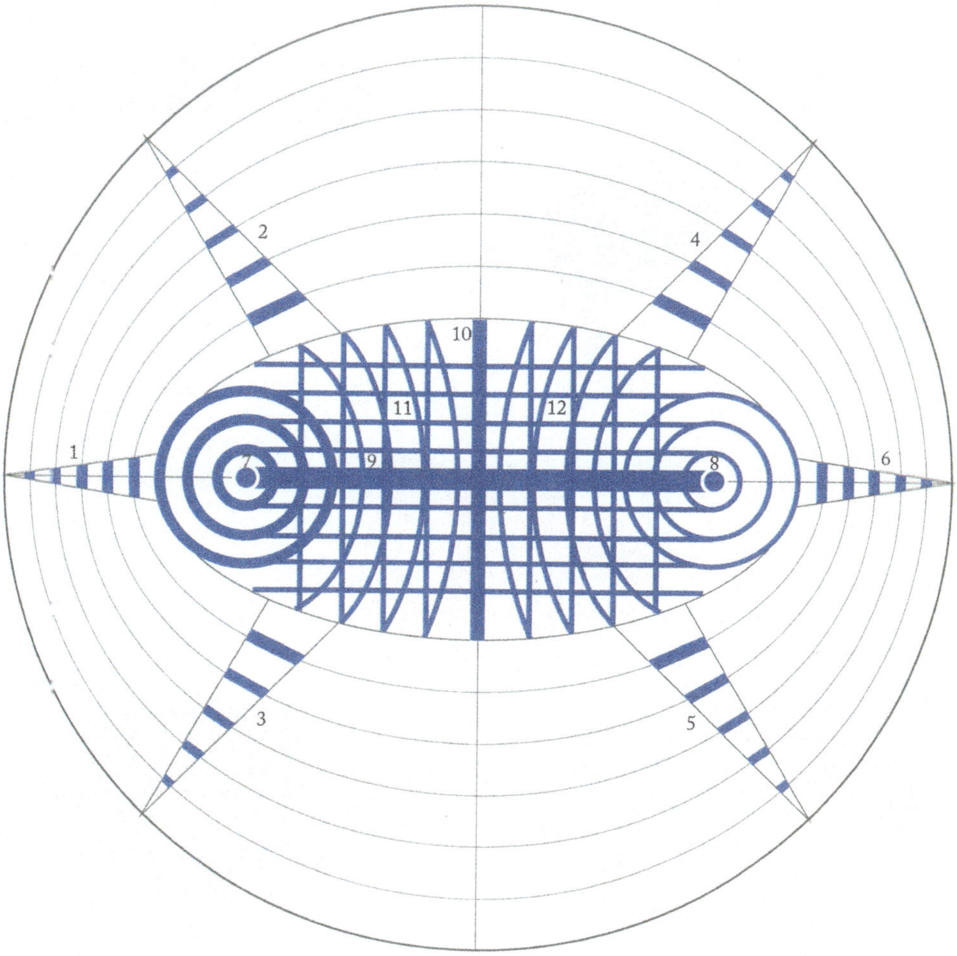

Fig. 11.6 The significant geometric traits of vertebrates combined in a single figure. The number of species in vertebrates is about 50,959 (24,500 fishes, 5,000 amphibians, 7,984 reptiles, 9,000 birds and 4,475 mammals) (Burnie 2004). Obviously, not all of them are depicted in encyclopedias or other treatises, but the most representative have been included in these works and are those used in the construction of the geometric disk. Its preparation is solely based on the traits displayed by the above mentioned species which was combined with the molecular and genetic evidence available. A detailed description is found in the text. The main features of the circular diagram are: **1 Bill**. In birds it has transversal stripes. The bill occurs as well in mammals (platypus) and bill-like structures are common in fish. **2** and **3 Forelegs** and **4** and **5 Hindlegs** are covered by *longitudinal stripes*. **6 Tail** with transversal broad stripes. **7 Eye, Primary Pole**. The ancestral veretebrate eye, on top of the head, is depicted instead of lateral eyes because it is the earliest form of photoreception in vertebrates. The *circularity* of the eye extends to the head and most parts of the body (best seen in birds). **8 Rump** or **Rear, Secondary Pole**. The *circularity* of the rump is displayed from humans to fishes. **9 Heavy Spine Band**. It runs along the vertebral column separating left from right side in a sharp way. **10 Line Dividing Anterior from Posterior Region**. It separates the two halves of the trunk (most evident in zebras and fishes). **11 Anterior Region of Body**. It is covered by *circular stripes* as well as by *parallel horizontal* and *vertical lines*. **12 Posterior Region of Body**. Is equally covered by *lines* with the same three orientations. All rights reserved covering this original Figure. This can only be reproduced with permission of the author or his heirs being accompanied by: (1) The complete text of the figure caption. (2) A citation as follows: Lima-de-Faria, A. 2014, "The Molecular Origins of Brain and Body Geometry. Plato's Concept of Reality is Reversed", to be accompanied by Figure Number, Page Number, Publisher, City, Country. (3) Consent will only be given for reproduction of the entire unmodified image

compass direction. Lizards learned to swim to the hidden goal and the intact parietal eye was found to be required to perform the orientation. When the parietal eyes were painted black the animals were completely disoriented (Beltrami et al. 2010).

It turns out that the third eye is physiologically and anatomically related to the pineal gland present in the lower part of the brain of most vertebrates. Pineal and pineal-related organs are present in the early vertebrates such as lamprey, teleosts and frogs. These organs can discriminate between different wavelengths of light. The same applies to the reptilian parietal eye which was found to sense UV light in *Iguana iguana* and *Anolis carolinensis* (Wada et al. 2012).

Fig. 11.7 (1) (2) and (3) **Bills and bill-like structures in birds (hornbill), mammals (platypus) and fishes.** (1) Wreathed Hornbill *Rhyticeros undulatus*. (2) Duck-billed Platypus *Ornithorhynchus anatinus*. (3) *Notopogon macrosolen*. **Fish species display stripe patterns characteristic of other vertebrates.** (4) *Acanthurus lineatus*. (5) *Pygoplites diacanthus*

11.8 Similarities and Differences Between Vertebrate and Crystal Geometries

Living organisms seem to have nothing in common with crystals, but like minerals they have inherent atomic order and their own geometry. The first is due to the self-assembling capacity of their main molecules such as proteins (Chap. 17). The second is decided by specific molecular cascades originating in their DNA sequences (Chap. 7).

Thus, they have basic properties in common with mineral crystals. But where lies the difference?

Amorphous materials that before were considered to have no relationship with crystals have turned out to be "shapeless, but not without order". They have a short-range order which represents the starting point in a transition process having at the other extreme the long-range order evident in minerals.

Quasicrystals, that earlier were considered impossible, were found to represent an intermediate situation which has poor long-range order but a high degree of short-range rotational symmetry. This new type of crystal has a "highly ordered atomic structure, yet the clusters repeat in an extraordinarily complex nonperiodic pattern". They are found in colloids and molten metals (Chap. 17).

The internal order of living organisms is maintained from generation to generation but shows at the same time diversification. In vertebrates the original mould—like the seed from which a crystal grows—could not be evaded. The skeleton of fishes can be recognized in the skeleton of the human body (Fig. 3.5), but there was a high freedom of combination of the cellular elements into novel alleys that obliged to separate living organisms into what are called: families, orders and classes.

Besides, like quasicrystals, living organisms have a geometry, but this is also more diversified. In quasicrystals

Fig. 11.8 **Fish species displaying in a distinct way the body figures seen in other vertebrates showing how the same molecular geometry is found across the whole group.** (**1**) Zebrias Regani *Synaptura regani*. (**2**) *Apogon quadrifasciatus*. (**3**) *Oplegnathus robinsoni*. (**4**) *Chromis dimidiata*. (**5**) *Pristigenys niphonia*

the fivefold symmetry dominates but in animals it can extend from twofold (human body) to thirteenfold symmetry (starfishes).

Additionally, the geometry of living organisms may be both periodic and nonperiodic. The patterns observed consist of lines which are parallel and distributed with a regular spacing but they may also be formed by areas without markings. Alternatives are permitted.

It seems that the difference lies essentially in an increase in the degrees of freedom which allow a larger combination of the atomic structures, being at the same time prisoners of a long-range coordination (Chap. 17).

Due to these properties, living organisms cannot be easily separated from other types of crystals. They demand, however, to be classified as *ultracrystals* forming a separate category.

Ultracrystals have several other properties in common with other crystal types. In minerals the long-range order leads to the subsequent addition of molecular layers as the crystal grows. Although the crystal increases in size it maintains the pattern of the original structure. The same process takes place in the carapaces of turtles. In them there is an addition of layers, which become successively larger but also maintain the pattern of the initial structure.

The long-range order of mineral crystals has, as well, its counterpart in invertebrates and vertebrates. It was described, since the early days of biological research, under the name of "regeneration". Flatworms, e.g. *Planaria maculata*, have a body orientation with a distinct head and a tail. If the animal is cut through the middle into two parts, the head section will increase in length, building by molecular processes a tail. The remaining tail section, following the same mechanism, will create a new head which has a brain and eyes. Two complete organisms are formed from the original separate parts (Morgan 1934). Regeneration is actually an initial form of reproduction which follows simple long-range copying.

Vertebrates are also able to perform the same feat but in a limited way: lizards build a new tail when this is severed and frogs can form additional legs (Halliday and Adler 2004).

Hence, invertebrates and vertebrates can use the long-range order earlier present in minerals.

At the origin of the cell, this order assumed the form of what is called "reproduction" in which the long-range information is transmitted with the help of specific organelles called "chromosomes". They carry the imprint of the whole pattern in a more precise way since it is written on their DNA sequences. But let us not forget that DNA only initiates this event. A whole series of molecular levels: RNA transcription, protein translation and molecular reshaping, occur before the final pattern is obtained. The various levels are a process of long-range action in which each molecule delivers to the next one its initial information.

This transfer is highly guided. If faults arise in the long chain there are repair mechanisms on every level. There is DNA repair, RNA repair and even protein repair (Cakmak 2009).

Hence, living organisms are not essentially different from mineral crystals. Amorphous materials represent the initial form, in a series of transition states, that led at the other extreme to the formation of ultracrystals—the living organisms.

Sources of Figures

Fig. 11.1 (**1**) Bramwell, M. (Editor) 1980. The Atlas of World Wildlife. Mitchell Beazley, London, UK (Fig. page 116). (**2**) Pope, J. 1986. Do Animals Dream? Michael Joseph, London, UK (Fig. page 46).

Fig. 11.2 (**1**) Wilson, D.E. and Mittermeier, R.A. (Editors) 2009. Handbook of the Mammals of the World, Vol. 1. Lynx Edicions, Barcelona, Spain (Vol. 1, Fig. page 124, photograph). (**2**) Wilson, D.E. and Mittermeier, R.A. (Editors) 2011. Handbook of the Mammals of the World, Vol. 2. Lynx Edicions, Barcelona, Spain (Vol. 2, Fig. page 110, photograph). (**3**) and (**4**) del Hoyo, J. et al. (Editors) 1996. Handbook of the Birds of the World, Vol. 3. Lynx Edicions, Barcelona, Spain (Vol. 3, Plate 33, Nr. 6 and Plate 38, Nr. 41).

Fig. 11.3 (**1**) and (**2**) Wilson, D.E. and Mittermeier, R.A. (Editors) 2011. Handbook of the Mammals of the World, Vol. 2. Lynx Edicions, Barcelona, Spain (Fig. Vol. 2, Plate 38, Nr. 131). (**3**) Wilson, D.E. and Mittermeier, R.A. (Editors) 2011. Handbook of the Mammals of the World, Vol. 2. Lynx Edicions, Barcelona, Spain (Fig. Vol. 2, Plate 35, Nr. 110). (**4**) del Hoyo, J. et al. (Editors) 2002. Handbook of the Birds of the World, Vol. 7. Lynx Edicions, Barcelona, Spain (Vol. 7, Fig. F1, page 18). (**5**) Smith, M.M. and Heemstra, P.C. (Editors) 1986. Smiths' Sea Fishes. Springer, Berlin [Fig. Plate 73, Nr. 204.9(A)].

Fig. 11.4 (**1**) Broman, I. 1933. Lärobok i Människans Anatomi. Gleerups, Lund, Sweden (Fig. 37, page 39). (**2**) and (**3**) Wilson, D.E. and Mittermeier, R.A. (Editors) 2011. Handbook of the Mammals of the World, Vol. 2. Lynx Edicions, Barcelona, Spain (Fig. Vol. 2, Plate 5, Nr. 6 and photo page 119).

Fig. 11.5 Lima-de-Faria, A. 2012. Molecular Geometry of Body Pattern in Birds. Springer, Berlin, London, UK (Fig. 8.36, page 112).

Fig. 11.6 Original. Graphic representation of body pattern geometry in vertebrates. Not to be reproduced without complete reference to this work.

Fig. 11.7 (**1**) del Hoyo, J. et al. (Editors) 2001. Handbook of the Birds of the World, Vol. 6. Lynx Edicions, Barcelona, Spain (Vol. 6, Plate 42, Nr. 46). (**2**) WIKIMEDIA.org/WIKIPEDIA, commons. (**3**), (**4**) and (**5**) Smith, M.M. and Heemstra, P.C. (Editors) 1986. Smiths' Sea Fishes. Springer, Berlin (Fig. Plate 24, Nr. 147.4; Plate 33, Nr. 610 and Plate 72, Nr. 204.13).

Fig. 11.8 (**1**) Smith, J.L.B. 1949. The Sea Fishes of Southern Africa. Central News Agency Ltd., South Africa (Fig. page 156, Plate 10, Nr. 320). (**2**) Smith, J.L.B. 1949. The Sea Fishes of Southern Africa. Central News Agency Ltd., South Africa (Fig. 485, Plate 23). (**3**), (**4**) and (**5**) Smith, M.M. and Heemstra, P.C. (Editors) 1986. Smiths' Sea Fishes. Springer, Berlin (Plate 78, Nr. 206.3; Plate 86, Nr. 219.14 and Plate 47, Nr. 174.4). Most of the figures of fishes included in this work were painted by artists having the initials: *MMS*, *DPV* and *HWBM*.

Part III

Geometry: An Obsession of the Animal and Human Minds

12 Science from Its Onset to the Present Has Been Pervaded by Geometry

12.1 The Fundamental Simplicity of Geometry Is Demonstrated by the Origin of Its Curves

One hundred years after Euclid, the Greek mathematician Apollonius (c. 260–190 BC) of Perga (a Hellenistic city in Anatolia) wrote a treatise "On Conic Sections". In it he showed that all major curves that are part of geometry can be derived from two cones being intersected by a single plane having different orientations. The resulting geometric figures are the circle, ellipse, parabola and hyperbola. This most original work represents the high peak of Greek geometry. His work demonstrated that a fundamental simplicity pervades geometry (Fig. 12.1).

Apollonius was also interested in astronomy and especially in the moon and proposed a theory on the motion of planets. Much later Kepler arrived at the conclusion that planets moved in ellipses (Ian et al. 2002).

Contemporary with Apollonious was the finest scientist and mathematician of the ancient world Archimedes of Syracuse (c. 287–212 BC). He used geometrical methods to measure rigorously curves, and the areas and volumes of solids, such as the volume of the sphere. By applying mathematics he invented the water-screw for moving water up a slope. Later the Greek–Roman writer Plutarch (c. AD 50–c. 120) summarized the obsession with geometry that pervaded the mind of scientists by stating that: "Archimedes asked his friends and relatives to place on his tomb after his death nothing more than the shape of a cylinder enclosing a sphere, with an inscription explaining the ratio by which the containing solid exceeds the contained" (Plutarch 1965).

12.2 It Is Not by Accident that the Sudanese and Egyptian Pyramids Remained Nearly Intact for over 4,000 Years

The great size and shape of the Sudanese and Egyptian pyramids has impressed generations of observers (Cheers 2003). The question is seldom asked: how could they have endured the destructive force of erosion over the millennia, being built stone on stone without cement. The answer lies in the profound knowledge of mathematics and especially geometry that these two African countries developed before any one else. Later Greek and Roman science only improved on these original discoveries. The over 80 pyramids found in these two countries are "astonishing works of engineering" (North 2008).

There was "an organizing principle common to ancient pyramids, sculptures and drawings" (Romer 2007). As he explains, the pyramid's interior architecture was set in the manner of a sculpted relief within a narrow envelope of space. Like a sculpted drawing or relief, it was set up upon a grid of squares (Fig. 12.2).

The internal architecture of the Great Egyptian pyramid was based on a six-square grid. The exterior was conceived as a sculptor's block whose four vertical planes had been pushed together into a single point. Also a set of co-ordinates that controlled the rising angle of the pyramid were employed. The Great Pyramid has been studied most extensively and was found, in addition to the grid of squares, to have its geometry guided by two staggered grids of equal size and on the same vertical planes. This is a genuine abstract principle as demonstrated by the different angles that they form. Romer (2007) concludes that the stability of the stone construction is due to this geometric design.

A. Lima-de-Faria, *Molecular Origins of Brain and Body Geometry*,
DOI 10.1007/978-3-319-06056-9_12, © Springer International Publishing Switzerland 2014

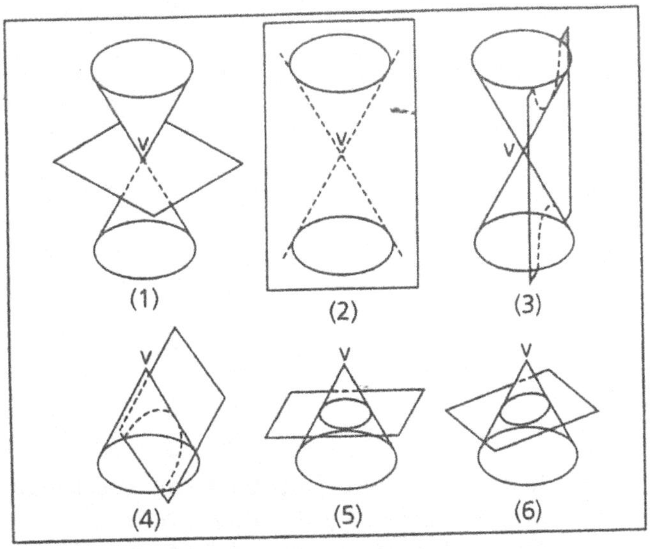

Fig. 12.1 **The simple origin of the major geometric curves.** These are obtained as conic sections. A *cone* (sometimes double) is cut by a plane to give: (**1**) a single point, (**2**) a pair of *straight lines*, (**3**) a *hyperbola*, (**4**) a *parabola*, (**5**) a *circle* and (**6**) an *ellipse*

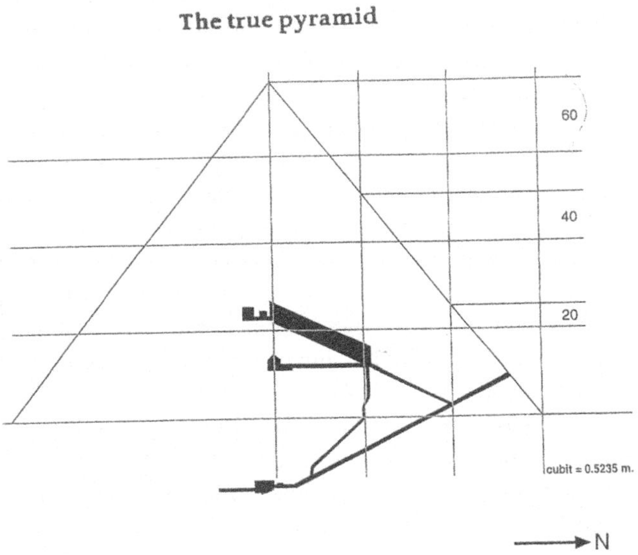

Fig. 12.2 The true *pyramid* was built using the Egyptian grid system

12.3 The Evolution of the Pyramids: The Human Mind in Search of Perfect Geometric Solids

Science is a dynamic process characterized by the inclusion of novel solutions. Mathematics and architecture evolved in Africa as elsewhere during the early millennia of human civilization.

Initially pyramids had a different form, size, and interior design. Between 2,630 and 1,640 BC, kings ordered the building of their tombs in the form of pyramids. Two main types were raised: the step and the true pyramid. The earliest pyramids, dating to the 3rd Dynasty consist of several "steps" and the burial chamber was situated below ground level. One of them is located at Saqqara and belonged to King N. Djoser.

The true pyramid was introduced at the beginning of the 4th Dynasty. It was a natural development of the earlier step pyramid, due to the advancement in mathematics that had occurred in the mean time. The burial chamber was now located above ground and the whole construction was accompanied by minor adjacent pyramids enclosed within a wall (Fig. 12.2).

The geometrical preoccupation of African scholars led to the subsequent building of monuments called obelisks. These are actually gigantic prisms. One of the best known was raised by Rameses II outside the Temple of Amun-Kamutef at Luxor. From the beginning they were a pair, but one of them was transported to Paris where it stands at the Place de la Concorde.

12.4 The African Square Grid Was an Invention that Has Since Then Been an Essential Element in European Thought

The wheel was discovered about 5,000 years ago. It has hardly been improved upon since then and it remains the essential component of most moving instruments.

The square grid dates from the same time. It has been used since then in western civilization until modern times as an essential component not only in drawing and sculpture but also in science.

The grid started to be used in the production of reliefs and paintings, by drawing guide lines, that later became square grids. These also facilitated copying. Earlier grids were based on 18 squares from the ground to the upper part of the head when representing the human figure. This also allowed the enlargement of a design mechanically from a smaller draft. Grids were also used for the representation of animals such as the bull which was often depicted and for the construction of the pyramids (Fig. 12.2).

The combination of biology with mathematics has led to the extensive use of square grids throughout the centuries to elucidate the changes in shape of biological form.

12.5 Greeks Were Highly Preoccupied with Mathematics and Its Relationship to Biological Form

During the 1900s an idea dominated in Western culture. It was taken for granted that the Ancient Greeks had discovered most of our basic ideas and fundamental principles. Since then this approach has been superseded by archeological and historical research revealing that most of these concepts had developed much earlier in Babylonian society, the African countries, India and the Far East.

One example is the "Pythagorean Theorem" which states an equality between the lengths of the sides of a right-angled triangle: the length of the hypotenuse squared is equal to the sum of the squares of the two other sides. This relationship was known long before Euclid and Pythagoras. It already appeared in a Babylonian cuneiform tablet dated to 1800 BC. This rule was also known in ancient India and Egypt (Crease 2009).

What later makes Greek thought so distinctive from other societies in the Mediterranean basin, is the extreme preoccupation with the significance of numbers and their relationship to music and biological form.

Pythagoras was born circa 570 BC. He emigrated from Samos to southern Italy in c. 531 where he was active. For him all things could be expressed numerically. He insisted that the universe could be reduced to mathematical terms thereby initiating the arithmetical interpretation of nature (Blackburn 1996).

This approach led to the discovery of the basis of acoustics—the numerical ratios underlying the musical scale. The way he found this relationship was solely experimental. While passing the workshop of blacksmiths he overheard the beating of hammers. Somehow these emitted a single consonance from differing sounds. He decided that the strength of the men hammering caused the diversity of sounds. Examining the weight of the hammers he recognized that those sounding together the consonance of the octave were double in weight. Further he discovered how different weights led to mathematical ratios resulting in specific musical notes (Grout and Palisca 2001). A lyre, for instance, could not be used to play an acceptable melody when its strings were tuned at random. It was further established that there was a relationship between the measurable lengths of the chords of a lyre and audible harmony.

From these experiments it was inferred that musical relations correspond to intelligible mathematical principles and since music is a paradigm of harmonius order the Greeks sought to describe the ordering of nature in terms of similar numerical relations and principles (Hornblower and Spawforth 1999). This way of thinking soon extended to scientific disciplines such as biology. Aristotle (384–322) followed this need to unify nature by systematizing the animal and plant world for the first time.

Art also became incorporated into the ordering of the universe, being approached like science. Sculptures of the human body had to follow a rigid canon (a word that means rule or law in Greek). Proportional relationships, called by the Greeks *symmetria* were of fundamental importance, especially for statues (Honour and Fleming 2002). The beauty of a human figure was considered to arise in the commensurability of the parts such as finger to finger, of the fingers to the palm, of these to the arm and to the whole body building a unity. This is why a statue had to be completely nude to display these relationships properly. Only in later periods did the male and female bodies become partly covered as in the Venus de Milo (which belongs to the Hellenistic period c. 330 BC). Science demanded nudity.

12.6 Euclid's Geometry Influenced Our Way of Thinking Throughout the Ages: Leonardo Da Vinci and Dürer As Scientists

Euclid was a mathematician who lived between 325 and 250 BC and who wrote a textbook of elementary mathematics, called "Elements". It has remained the standard work for 2,000 years of teaching geometry. Attributed to Euclid are also two treatises on music "Section of the Canon" and "Harmonic" showing the connection between geometry and music.

One of the many editions and translations of the "Elements", was published in 1847 by Oliver Byrne, in which the figures were depicted in colors to allow students to grasp better the various theorems. Euclid's work had a great impact on Roman science. It influenced Pliny the Elder (AD 23–79), a prominent Roman, who is best known for a 37 volume encyclopaedia of all contemporary knowledge of animals, plants and minerals.

Architecture was also strongly influenced by Euclid's geometry. It is not by accident that the large Roman theatres stand after 2,000 years. They were built in such a way that they are still used for stage performances and their acoustics are unrivalled. Other buildings like huge aqueducts, at Nimes in France and Segovia in Spain, stand undisturbed and one of them continues to furnish water to the city as it has done throughout the centuries.

The way these architectural masterpieces were built is described in detail by the Roman architect Vitruvius, who served the emperors Caesar and Augustus, in his treatise "De Architectura". As he acknowledges he is heavily indebted to Greek authors.

Vitruvius' importance as an engineer and architect extended to his treatment of science ranging from mathematics to medicine. According to him architecture should include everything which touches on the physical and intellectual life of man (Hornblower and Spawforth 1999). Such a unitary concept had great influence during the Italian Renaissance 1400s to 1500s. His work became a permanent source of reference in which human anatomy and Euclid's geometry were combined. Humans had not been depicted nude for over 1,000 years and their anatomy had been a forbidden area of research.

One of the leading exponents of this period, as a scientist and artist, was Leonardo da Vinci (1452–1519) who re-interpreted the ancient teachings of Vitruvius on the

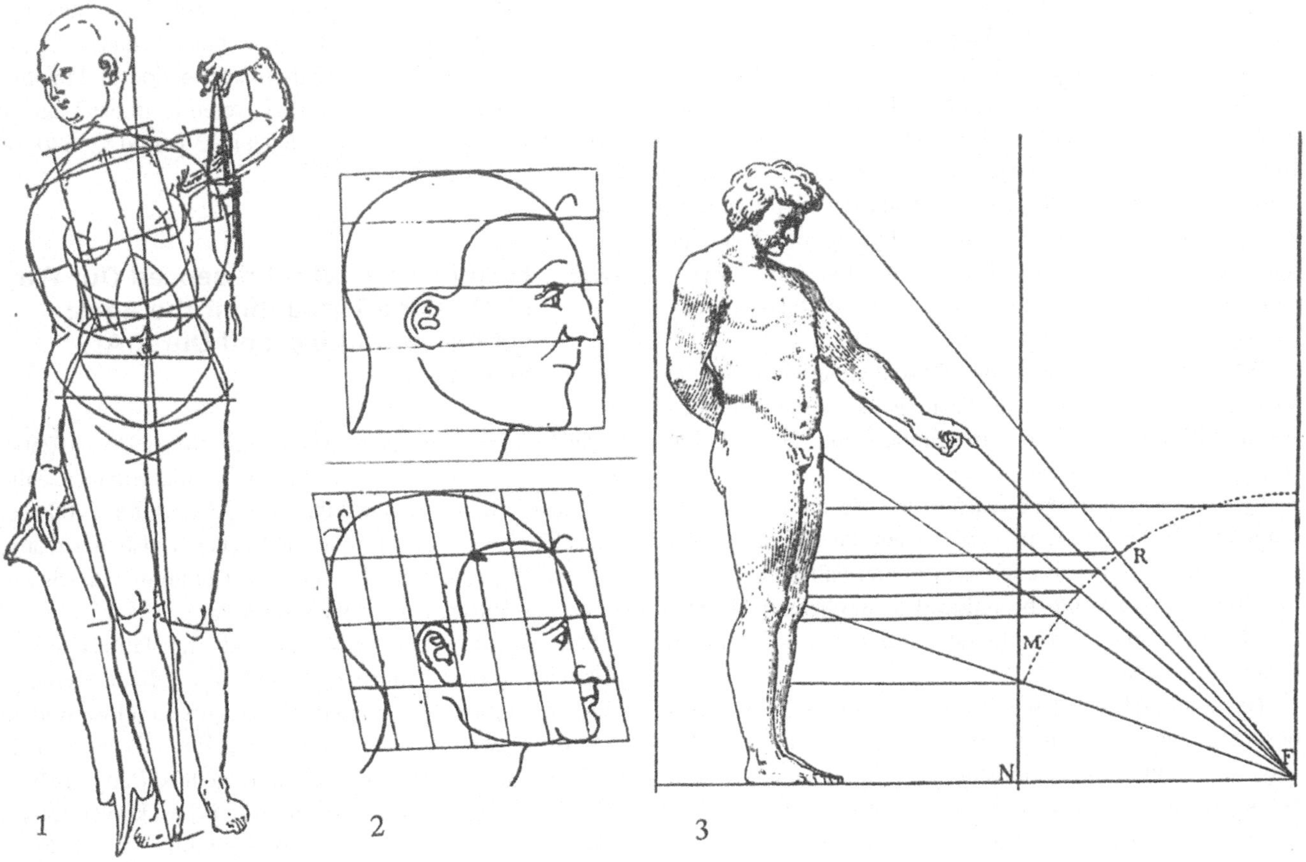

Fig. 12.3 Renaissance scientists and artists showed how the different shapes of the same structure could be understood by transformations of the original grid. (**1**) The female body divided into *concentric circles* having their centers in the nipples and breast (Dürer 1506). (**2**) The Egyptian grid used by Dürer to derive the different shapes of the human head from an initial type. (**3**) Drawing by Leonardo da Vinci showing the division of the human body into geometric figures. These concretize the need, during the Renaissance, to reduce living organisms to geometric patterns

proportions of the human and animal bodies (Fig. 12.3). As the Renaissance movement expanded into Germany, Albrecht Dürer (1471–1528) not only followed Euclid and Vitruvius but also used the Egyptian square grid to represent the different shapes of the human head (Fig. 12.3).

12.7 The Search for Geometric Proportions in Astronomy

With the advent of the 1600s the Renaissance movement was left behind. The great expansion in overseas trade led to a rapid development of experimental science. Galileo Galilei (1564–1642) in Italy, used the telescope for the first time to discover that the surface of the moon had mountains and valleys like those of the earth, and to find out that Jupiter had its moons like our planet. Besides he created experimental physics by establishing the laws of mechanics. His contribution to the subsequent growth of physics and astronomy was paramount. It was his ground work that allowed Kepler and Newton to later build their theories.

Kepler (1571–1630), who discovered the laws of planetary motion, was highly preoccupied with the geometry of the solar system. His results were the product of long and painstaking work. Being a leading mathematician, Kepler tried from the beginning to see if the orbits of the planets fit into circles, triangles and other geometric figures. It was only after 10 years of careful calculations that he came to the conclusion that their orbits had to be ellipses and not circles as previously thought. Moreover, the sun was located at the focus of the ellipse that the planet described during its rotation. As Padamsee (2003) points out: "Kepler had fulfilled Pythagoras' prophecy. There was a mathematical pattern governing planetary orbit sizes. Indeed, it surpassed the Greek's original ambition. Kepler's Third Law unified orbital sizes and speeds."

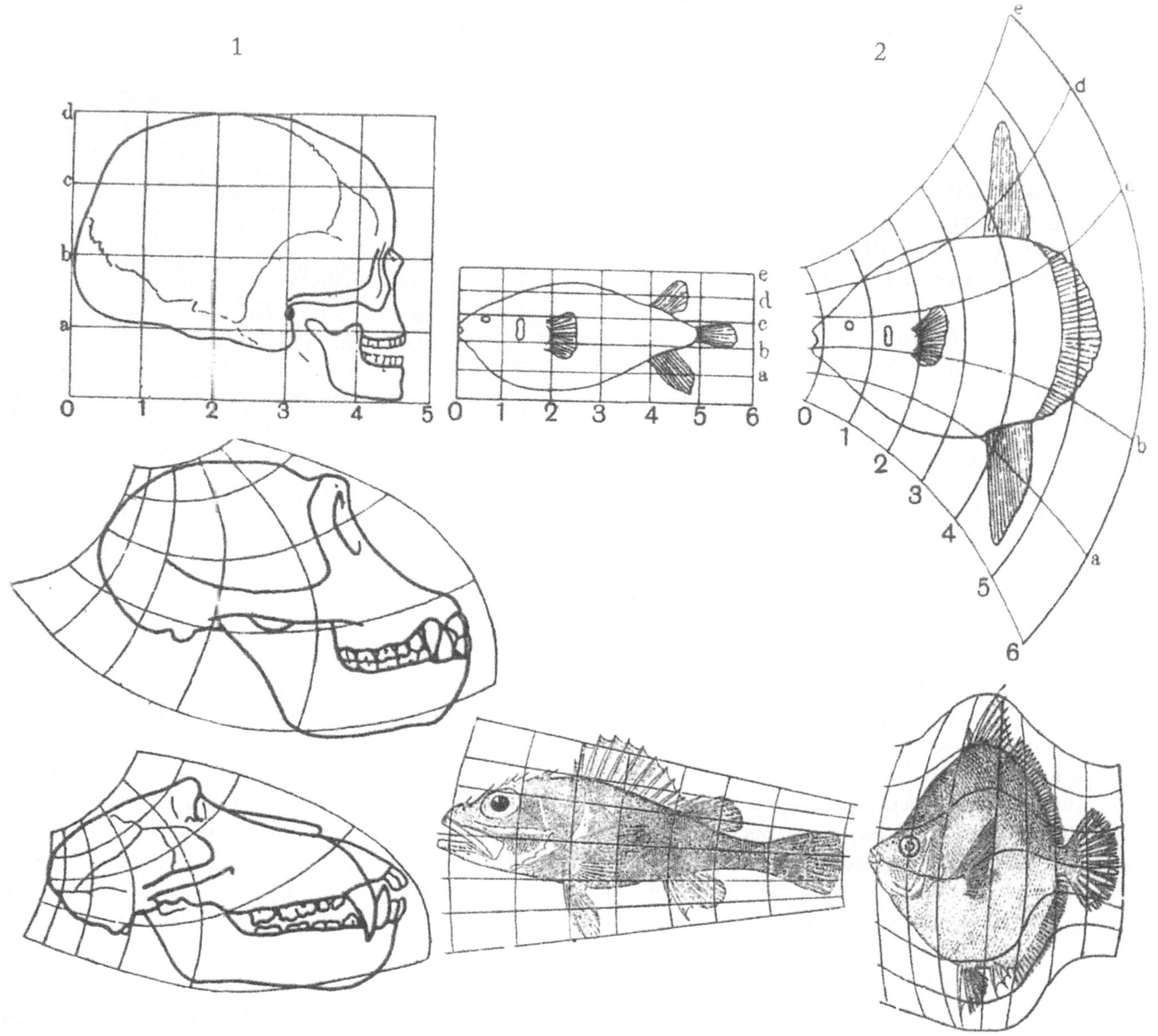

Fig. 12.4 Following the efforts of earlier scientists, the Egyptian grid was extended in the 1900s to quite different living organisms to account for their body transformations. (**1**) Human skull covered by the Egyptian square grid, and the skulls of a chimpanzee and baboon derived from it, following Dürer's transformation principle. (**2**) Transformations in fish body shape using the grid (from Thompson 1917)

12.8 The Egyptian Grid Entered Modern Biology to Explain the Transformation of Animal Form

The naturalists of the 1600s and 1700s were mainly preoccupied in describing the large number of animals and plants which they encountered during their voyages of exploration around the globe. It was only by the beginning of the 1900s that biology became an experimental science and the geometry of living organisms could be initiated. D'Arcy Thompson went directly to the work of Dürer which he includes in his book. Dürer had used the Egyptian grid, and its transformations, to represent the changes in shape of the human head.

In his seminal work "On Growth and Form" Thompson (1917 reprinted 1952) analysed biological processes from their mathematical and physical aspects explaining biological growth and form as a unitary process in which the form of one species could be shown to be derived from a previous one following a geometric transformation (Fig. 12.4). His work that is widely recognized as a classic has been ignored to a large extent, and continues to be mentioned only as a side event, because he refused to follow the Darwinian interpretation of evolution that still dominates.

12.9 Euclidean and Non-Euclidean Geometries in Present Day Physics and Biology: The Difference Between a Flat and a Curved Triangle

It is not by accident that D'Arcy Thompson's work was published 2 years after Albert Einstein (1879–1955) developed his general theory of relativity in 1915. This was a period of intellectual ferment when most accepted scientific ideas were challenged.

Einstein had found that his relativity concept required that the space of the universe be considered a non-Euclidean space. It needed another type of geometry which had been devised previously by the mathematician Georg Riemann (1826–1866) who changed the concepts of distance and curvature into a multidimensional geometry. Einstein created a non-Euclidean geometry in which curved space and time were the novel dimensions.

Since then non-Euclidean geometry has become an independent branch of science. The original Euclidean geometry deals mainly with geometric figures drawn on a flat two-dimensional surface. When the circles and triangles are laid on the surface of the earth, they do not obey the flat Euclidean geometry. Pythagoras' theorem on the hypotenuse of a right-angled triangle does not hold and the sum of the interior angles of a triangle on the surface of a sphere is more than 180° (Fig. 12.5).

Physical and chemical processes regulate the structural organization and function of biological structures. The present development of molecular biology has shown how the geometric patterns displayed by the body of birds and other vertebrates are the result of specific gene actions leading to the formation of proteins and other molecules that guide the pigments to their exact geometric locations. This biological geometry which is decided by molecular processes has been designated as molecular geometry (Lima-de-Faria 2012). This is a non-Euclidean geometry because the geometric figures of living organisms are not laid on a flat surface, but instead are formed on the curved shape of the animal's body. Besides this geometry is characterized by having three more dimensions. These are: (1) The transmission to the progeny of the same basic pattern by means of a chemically organized structure, the chromosome. (2) The temporal development of a fertilized cell leading to an embryo and later to a full grown organism, events which are directed by specific macromolecules such as messenger RNAs and microRNAs. (3) The modification of the animal's body structure and function by chemical and physical agents of internal and external origin (e.g. hormones and radiation). To be noted is that the molecular geometry found in an animal's body is ephemeral, since it changes with age and disappears with the death of the individual. Yet it persists because, as a consequence of being the product of ordered genetic processes, it returns untouched in the next generation.

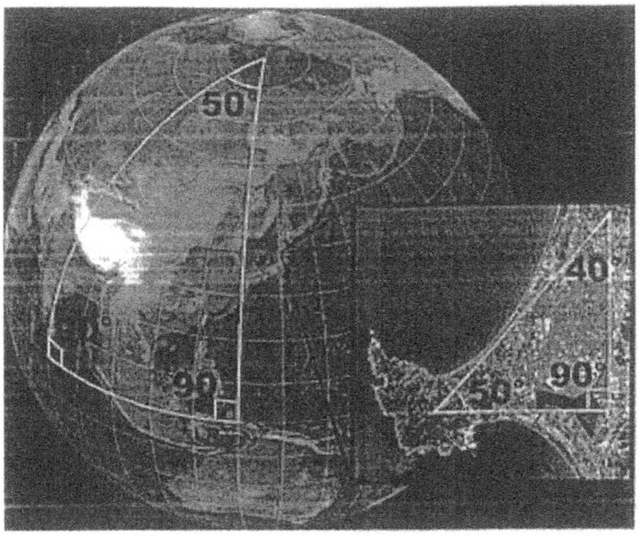

Fig. 12.5 Euclidean and non-Euclidean geometries. Non-Euclidean geometry. A *sphere* showing a *triangle* built on its surface with angles of 90°, 90° and 50°; compared with (*to the right*) Euclidean geometry. A *triangle* on a flat surface with angles of 90°, 50° and 40°. On a *sphere*, the sum of the angles of a *triangle* is not equal to 180°

12.10 A Tortoise Is a Lesson in Non-Euclidean Geometry: The Triangles and Hexagons of Its Shell Obey Non-Euclidean Geometry Because They Are Curved

The Radiated Tortoise *Asterochelys radiata* is considered "One of the most beautifully marked tortoises" (Cogger and Zweifel 1998). Its carapace which has a shape close to half a sphere is covered with triangles and other geometrical figures, making it a lesson in non-Euclidean geometry since the triangles are on a curved surface (Fig. 12.6). Another species, *Gopherus polyphemus*, makes this geometry still more palpable with its upper side consisting of hexagons and pentagons, taken out of a mathematician's book, but these are also non-Euclidean since they form a partly spherical structure. The ventral body shield of this species is also

1

2

Fig. 12.6 Molecular geometry in one of its most patent forms. Tortoises exhibit startling geometric patterns that are non-Euclidean. (1) "One of the most beautifully marked tortoise", radiated tortoise, *Asterochelys radiata*. (2) The African turtle *Testudo radiata*. The *triangles*, and other geometric figures on their shells, are not flat Euclidean *triangles* but are located on a *spheroid* surface being, as a consequence, non-Euclidean

highly geometrical. It is covered by parallel squares which increase successively in size.

We tend to think that the final pattern is a direct product of the DNA message. But this is true only to a certain extent. The initial messenger RNA copied from DNA transmits its message to a protein that gets a specific amino acid configuration. It is however, the single atoms that the protein, or other resulting molecule, presents to other cell components, that are critical in the final decisions of pattern formation. It is their intervention during embryonic development that decides the final shape of the organism. These single atoms are the same that exist in minerals and which also determine their final form. It is thus not surprising that the geometric figures found in the shields of tortoises, and in other animal body configurations, are quite similar to those of crystals, sometimes even undistinguishable from them. DNA is most important in carrying the initial genetic message, that is the main determinant of the development, but it is far from being the sole director. It is the many intermediate atomic states that finally decide the final structure of the adult organism.

12.11 Guidance of Pigments and Brain Intervention in the Pigmentation of the Turtle's Shell: Turtles Change Their Color According to the Environment

The shell of a turtle is considered a remarkable structure because it is the product of an elaborate architecture (Halliday and Adler 2004). Both the outer layer of the skin and the inner tissue participate in the construction. The horny scutes on the surface consist of the protein keratin, associated with a layer of living cells. The deeper, thicker cell layer is formed of bony dermal plates. The dermis also contains the majority of pigment cells—the melanophores. These are loaded with black, yellow, white and red pigments (Kelsh et al. 2009).

The term melanophore derives from the fact that these cells mainly contain melanin. This pigment can be dispersed or concentrated leading to changes in color and body pattern. These transformations are known to be mediated by the nervous activity of the brain and by hormones secreted by glands such as the pituitary, and a combination of both sources.

The upper part of the turtle's shell, the plastron, may change its pattern due to environmental influences. During egg incubation, carbon dioxide concentration and temperature affect pigmentation in plastral pattern (Etchberger et al. 1993). Turtles reared on either black or white substrates for 120 days changed the coloration of their shells according to the background (Rowe et al. 2006).

Hence, the pattern of the shell is both constant and modifiable. On one hand it repeats itself from generation to generation with great accuracy being the basis of species identification, but it can also change within limits due to environmental factors. Rigidity and plasticity are not incompatible.

The black, brown and yellow pigments which are part of the triangles and pentagons and other geometric configurations on the shell of the tortoise *Asterochelys radiata* did not attain these positions by random processes. They were guided by the action of small molecules such as the melanocyte-stimulating hormone (MSH) a peptide which increases melanin synthesis in melanophores leading to skin darkening (Eckert and Randall 1978).

The configurations built by yellow pigments against a background of brown-black coloration, can now be understood in terms of a dynamic chemical process of molecular migration to specific sites on the carapace.

Fig. 12.7 The glass *pyramid* at the Louvre Museum in Paris built in 1984 by I.M. Pei. The obsession with *geometric solids*, as symbols of order and simplicity, continue to this day

12.12 The Obsession of the Human Mind with Geometry Extended into Modern Art: Cubism

The intellectual upheaval that led, in the early 1900s to the development of novel theories in physics, and resulted in a radical experimental approach in biology, extended its influence into art.

At the same time that relativity and quantum physics emerged (1905–1915) a parallel revolutionary approach took place in art. Paul Cezanne (1839–1906) became a pioneer in the interpretation of the shapes of fruits, and other biological structures based on an internal geometry. With his mastery he influenced younger generations of painters.

Pablo Picasso (1881–1973) was not satisfied with this approach. He wanted to go deeper. He reduced every structure, even the human figure, to the most elemental of all regular geometric solids—the cube (1910). Soon his novel interpretation of nature was followed by other painters creating an art movement known today as cubism.

It makes no difference whether we are dealing with science or art, our mind searches unremittingly for simplicity and coherence—geometry turns out to be an inborn obsession.

It is thus not surprising that the architect I.M. Pei, when asked by the French President François Mitterrand, to design a new entrance to the classic architecture of the Louvre Museum could not think of any other structure but a monumental glass pyramid (Fig. 12.7).

The cycle was closed: the Egyptian need for geometry resurfaced in the mind of twentieth century Europeans.

Sources of Figures

Fig. 12.1 Ian, D. et al. 2002. The Cambridge Dictionary of Scientists. Cambridge University Press, Cambridge, UK (Fig. page 9).

Fig. 12.2 Romer, J 2007. The Great Pyramid. Cambridge University Press (Fig. 30, page 65).

Fig. 12.3 (**1**) Waetzoldt, W. 1938. Tu y el Arte. Editorial Labor, Barcelona, Spain (Fig. 13, page 21). (**2**) Thompson, d'Arcy 1952. On Growth and Form, Vol. 2. Cambridge University Press, Cambridge, UK (Fig. 508, page 1053 and Fig. 510, page 1054). (**3**) da Vinci, Leonardo 1945. El Tratado de la Pintura. Spanish Version of the Edition of 1828, Buenos Aires, Argentina (Plate XVI).

Fig. 12.4 (**1**) Thompson, d'Arcy 1952. On Growth and Form, Vol. 2. Cambridge University Press, Cambridge, UK (Fig. 548, page 1082 and Figs. 550 and 551, page 1083. (**2**) Thompson, d'Arcy 1952. On Growth and Form, Vol. 2. Cambridge University Press, Cambridge, UK (Figs. 525 and 526, page 1064 and Figs. 523 and 524, page 1063).

Fig. 12.5 http://en.wikipedia.org/wiki/ Non-Euclidean geometry. 2011-08-15. (Fig. page 4).

Fig. 12.6 (**1**) WIKIMEDIA.org/WIKIPEDIA/commons (page 1). (**2**) Cabrera, A. 1935. Historia Natural, Vol. 1. Instituto Gallach, Barcelona, Spain (Fig. page 397).

Fig. 12.7 Wikipedia. Louvre Pyramid. Design and Construction (Fig. page 1).

Spirals Produced by Inert Matter As Well As by Spiders and Humans

13

13.1 What Are Spirals and What They Represent

The classical example of a *spiral* is that *of Archimedes*. This spiral starts at the origin and makes a curve with three rounds in which the distances between the branches are the same. The radius $r\,(t)$ and the angle t are proportional, the equation being $r\,(t) = at$ (in which a is a constant).

Another spiral is the *logarithmic spiral* in which each line starting in the origin cuts the spiral with the same angle. It is defined by several equations.

There are other types of spirals, including the *Fibonacci spiral*, which is drawn by putting two small squares on top of each other, to which is added a sequence of growing squares counter clockwise. This is completed by drawing quarter circles inside the squares. This spiral is also defined by an equation that produces the well known sequence of Fibonacci numbers.

Three-dimensional spirals build helices such as the well known DNA molecule, but they may assume other shapes like the *conical helix*.

If asked what a spiral structure represents there is a straightforward answer. A spiral is a curve, that conveys a specific order in the arrangement of matter.

13.2 Before Spirals Were Formed in Living Organisms They Occurred in Inert Matter

Galaxies consist of molecular clouds which serve as stellar nurseries. For decades astronomers have photographed and followed carefully the spiral shape which characterizes several galaxies, one of them being *M33*. This is our nearest face-on galaxy with pronounced optical spiral arms, consisting of giant molecular clouds. When studied with polarized light, the optical arms turn out to be aligned with the molecular cloud magnetic field that builds the spiral. The spiral formation is a molecular event occurring in space being directed by magnetic fields (Li and Henning 2011).

For a long time biologists and mathematicians have been fascinated by the occurrence of spirals in plants and simple organisms. But there is nothing extraordinary about this emergence if we consider that living organisms consist of the same chemical elements that are found in galaxies and other stellar matter. Besides there is no indication of any other different elements occurring in outer-space. We are dealing with the same chemical bonds coupling identical atoms. It is thus not surprising that spirals are part of the patterns that emerge in flowers, shells and body parts of living organisms.

In this connection it may be pointed out that plants have no brain and the same is true of single cell organisms such as protozoa.

13.3 Elementary Memory and Reactivity Are Already Present in Living Organisms Before There Is a Nervous System

Unicellular organisms, such as protozoans, have no cellular structures comparable to the nervous system found in multicellular animals. Nonetheless they are capable of detecting stimuli and responding adaptively. Some, even exhibit a rudimentary form of memory. Eckert and Randall (1978) ask "How is this possible without a nervous system?"

When a *Paramecium* bumps into an obstacle, it swims backward. If it gets a stimulus from the posterior end it swims more rapidly. Experiments using intracellular microelectrodes revealed that electrical charges and calcium concentration are involved in these responses.

Another protozoa, *Stentor*, when mechanically disturbed, contracts its body with the aid of an internal system of microfilaments. It actually remembers this disturbance

A. Lima-de-Faria, *Molecular Origins of Brain and Body Geometry*,
DOI 10.1007/978-3-319-06056-9_13,

because the response decreases exponentially with the number of trials. Microelectrodes were also used in this experiment. Eckert and Randall (1978) conclude: "It is evident from these two examples that the surface membrane of a unicellular organism can perform the functions of sensory reception, simple integration of receptor signals, and control of the motor response and that a rather simple modification of membrane function can give rise to an elementary form of memory. Thus the rudiments of the sensory-neuromotor systems of metazoans are present already in the electrophysiological properties of the Protozoa. The similarity between the electrical properties of nerve cells and protozoa suggests that these properties are very general and very ancient. This lends support to the view that the evolution of nervous systems has taken place, primarily, by the development of increasingly sophisticated neural circuits rather than through changes in fundamental properties of nerve cells."

13.4 The Brain Has Had an Inconspicuous Origin: It Started from Sensory Cells on the Skin

We are so attached to the idea that our brain is such an exceptional organ that we tend to consider it as coming from nowhere or having an evolutionary origin that puts it apart. This is an antiquated view that recent studies in comparative anatomy and physiology have disposed of.

As Withers (1992) points out, the brain has had an evolution, like any other organ. The origin of the brain is also much more modest than we could imagine.

The organization of sensory receptors, neurons and effector cells—which build the nervous system—started in the skin. These cells were originally located in the epidermis of animals such as the coelenterates. Later they aggregated and migrated building the clusters of neurons that formed ganglia and subsequently small brains. In higher organisms the central nervous system developed into an anterior concentration of neural tissue (the large brain) that become connected with the longitudinal nerve cords. This differentiation occurred already in spiders.

13.5 The Minute Brain of Insects Allows Them to Have a Language and to Perform Social Tasks Considered Characteristic of Human Behavior

Insects (which are closely related to spiders) display a high mental ability. Bees, ants, and termites are social insects that have developed: (1) a highly organized society, (2) a well-defined language, (3) possess dialects, (4) decide the fate of their progeny by differential fertilization, (5) have an agriculture consisting of the cultivation of fungi, (6) raise other species such as aphids, and use their sugary products in a manner akin to humans raising cattle and collecting their milk. Yet, they are not directly related to mammals.

Besides, the brain of a human weighs on average 1,400 g and consists of at least 10 billion cells (Changeux 1985), while the brain of an insect weighs between 1 mg and 1 microgram. The whole body of the common ant *Lasius niger* weighs about 0.6 mg and the number of cells in its brain is thought to be about 10,000. The difference could hardly have been bigger.

The honey-bees are considered to be the only animals other than man that communicate by a symbolic language (Dreller and Kirchnert 1993). Moreover, bees and ants, like birds, are known to use celestial navigation. They follow with different degrees of accuracy the movements of the sun and its relationship to the horizon.

A termite's nest is a complex edifice. The central chimney is a ventilation shaft that helps to keep the temperature constant. Outside changes, of as much as 13 °C occurring between day and night, do not affect the inside temperature. Other compartments function as: foraging tunnels, food stores, chambers where fungi are cultivated, the royal cell and rooms for the young termites.

In birds migration has been studied extensively. They use as cues the position of the sun, the magnetic field, the landscape, and other sources of information. Since these parameters change with time, migration has turned out to demand a high mental ability which evaluates and corrects the novel situations that they face when migrating to a specific region. Consequently there has been an evolution in bird migration imposed by the need to find novel solutions (Boyle and Conway 2007).

Insects with their microscopic brains also migrate and with high precision. The best studied case is the North American monarch butterfly *Danaus plexippus*. Its autumn migration comprises a journey of several thousand kilometers to the Gulf of Mexico. With the onset of spring the butterflies return north towards the Great Lakes of the U.S.A. They migrate in masses of many thousands. "So predictable is this movement with respect to direction, time of year and the vast numbers involved that many towns along the monarchs' fly path celebrate their arrival with festivals". (Baker 1980).

Baleen whales, that are among the largest living animals, migrate 20,000 km between polar and tropical waters each year. They are well known for their high mental ability which includes communication by a variety of sounds. Hence, the size and complexity of the brain do not seem to be decisive factors in determining this form of behavior since it occurs, with high precision, in butterflies, birds and whales.

Few of us think of ourselves as migrants, but humans are also known to have moved to novel geographic areas for thousands of years, populating other continents and continue to do so at present.

The behavior of insects shows many features comparable to those found in human society. This is an indication that the two types of brain—insect and human—have the same basic properties despite a tremendous difference in size.

13.6 Geometry Is Already in the Minds of Octopuses and Spiders

Invertebrates, such as octopuses, have the capacity to learn to distinguish between different geometric figures and can perform tasks in a way comparable to mammals.

Octopuses can recognize by sight 29 different shapes. They can be trained to distinguish between a cross and a square and 27 intermediate forms. They were given rewards of food when they attacked one of the symbols. The experiments carried out by *Stuart Sutherland* (of the University of Sussex, UK) revealed that the trained octopuses could not only distinguish between a cross and a square but could respond to other shapes with intermediate configurations. The geometry became evident when it turned out that the animals tended to classify shapes along a dimension that seemed to depend upon the relation of outline to area (Wells 1983) (Fig. 13.1).

Octopuses have revealed another unexpected capacity. Like primates they can invent tools. By diving in the Indonesian sea Finn et al. (2009) discovered that the octopus *Amphioctopus marginatus* was able to collect empty coconut shells and to assemble their halves as a shelter. It also manipulated and arranged the shells so that the concave surfaces were uppermost.

Spiders—with a brain of a size close to that of insects—have an equally refined geometrical mind. They are even a step further, because they can draw their geometric figures by actually building them in space. As they produce their webs several types of spirals are constructed with the utmost accuracy.

13.7 The Formation of Spirals As a Product of the Animal's Mind: Spider Webs May Be Geometric or Irregular Depending on the Species

Naturalists studying animal behavior, were astonished by the formation of spider webs in the form of regular spirals. To start with the spiral web is not formed as a spiral directly. The spider follows different patterns until the final spiral is formed. A species that produces such a type of web is *Gasteracantha falcicornmis* (Preston-Mafham 1996).

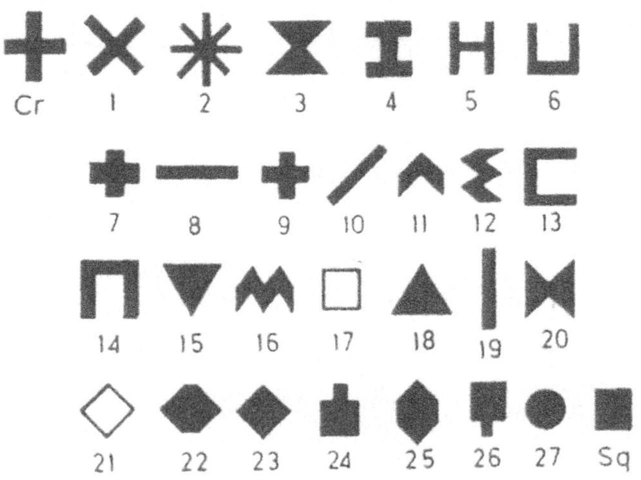

Fig. 13.1 **Geometry in the minds of octopuses.** Twenty-seven different signs plus a cross (*Cr*) and a *square* (*Sq*) can be recognized by octopuses. The numbers show the order in which the octopuses ranked the shapes. They were found to classify shapes according to a relation of outline to area

Surprisingly the spider starts by building a triangle, not a spiral. Hence, it must have a triangle in its mind before it concretizes it in space. A human cannot draw a triangle on a sheet of paper without previously having it in his (or her) mind.

The web in the form of a spiral is finished by following eight main stages. It starts as a triangle (stages 1 and 2) which is later transformed into a geometric solid close to a tetrahedron (stage 4). Then strong spokes unite at the center of a square (6). It is on this well-anchored frame that the temporary spiral is drawn (7). This is followed by the sticky spiral (8) (Fig. 13.2).

In other species there is a central hub followed by a strengthening zone and a free zone, but the main area of the web is covered by the sticky spiral.

Webs may take many forms depending on whether they are stationary or used as tools to trap insects. Spiders may also trap prey by jumping.

Foelix (1982), who studied spider behavior extensively, reached the conclusion that there has been an evolution in the building of the orb web. Two groups of spiders have formed simple radial webs but created also more complex structures like the spiral form. Some species, such as *Ixeuticus robustus* build irregular traps. Thus spiders use different solutions to obtain food. They build different types of tools that may take the form of spirals but may have a different geometry or no geometry at all (as when jumping). Their minds differ with the species to which they belong. The fact that the pattern of behavior is species-bound indicates that the geometric pattern has a genetic, i.e., a molecular basis.

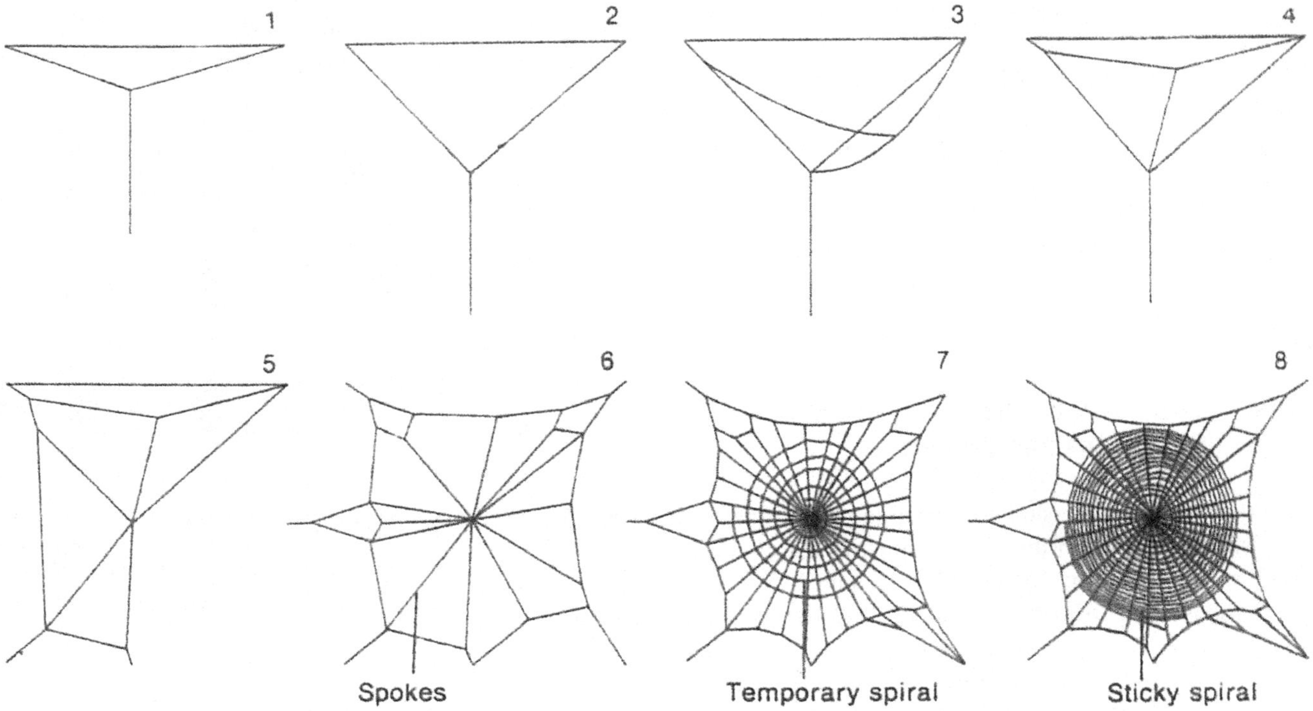

Fig. 13.2 Web construction by spiders. Spiral webs are the most familiar of spiders' traps. "First the spider makes a firmly anchored framework of strong, nonsticky silk. Spokelike lines are then made and held in position with a temporary, widely spaced spiral, later replaced with a closely spaced sticky spiral" (Pope 1986)

Experiments with drugs, using spiders in captivity, have shown that under their influence they are mentally disturbed, the result being the building of an inaccurate spiral. Age also affects this mental procedure. These results lend support to an explanation in which the spiral is conceived by the spider's mind before it is produced as a web. The situation has its parallel in humans, who under the effects of drugs or due to old age, also get their minds disrupted producing irregular writing and behaving in unpredictable ways.

13.8 The Same Neurotransmitters Are at Work in a Human and a Spider's Brain

The mind of a spider, located in its central nervous system, consists of nerve cells activated by brain chemical signals. Some are the same which occur in our brain. These are: (1) The amino acid *GABA*, which is restricted to the central nervous system and which is an inhibitory neurotransmitter. (2) Serotonin, which is a neurotransmitter is also present in the vertebrate brain. (3) Octopamine, also a neurotransmitter found in insects and vertebrates known for its role in behavior. As Barth (2002) points out there are even structural similarities between the spider's brain and the human brain. One of them is the distribution of blood vessels which follows an identical pattern in both structures.

13.9 Web Silk Genes Have Been Isolated and the Amino Acid Sequences of the Different Silk Proteins Were Determined Supporting the Single Origin of Orb Webs

Proteins are the product of genes which are segments of DNA found in almost every cell of an organism. Proteins evolve as do the organisms they build. Just as there are closely related organisms that are descendent from a common ancestor there are closely related proteins that are descendants of common ancestor proteins.

It had long been believed that the first spiders (about 220 million years ago) produced only one silk protein but present living spiders produce a number of different silks which they use in specific tasks: one silk cushions their eggs, another protects their egg cases, a third entraps their prey, a fourth builds the frame lines of the web and still another is employed to build the spiral lines (Brunetta and Craig 2010). Garb et al. (2006) concluded that "The spider orb web is an impressive example of animal architecture".

There are differences between the adhesive mechanisms of capture spirals spun by Araneoids (aqueous glue) and Deinopoids (dry fibrils). The two types of webs were considered as having occurred separately being an example of

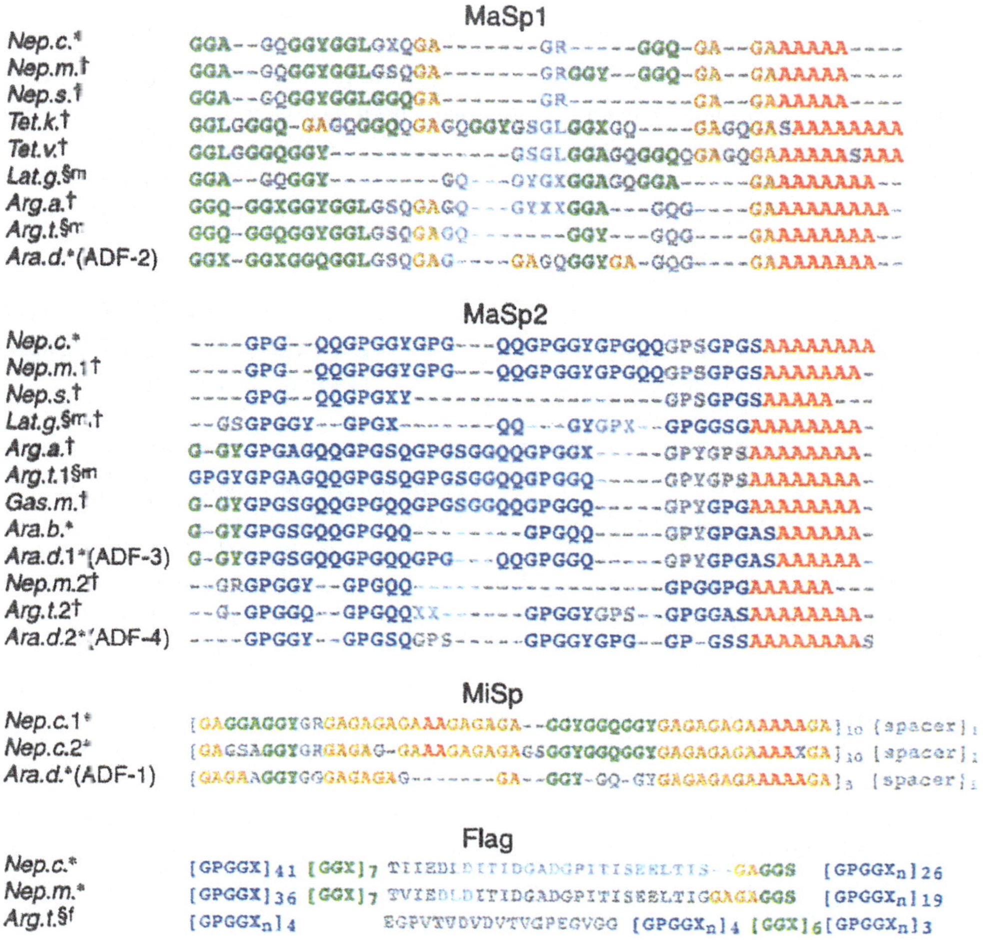

Fig. 13.3 **Sequencing of four araneoid silk fibroin proteins *MaSp1*, *MaSp2*, *MiSp* and *Flag*,** showing the repeat units of amino acids (*single letter symbols in color*) and their conservation in the different species (abbreviated names in the *left column*)

what is called convergence. However, the analysis of silk genes has led to another conclusion. Complementary DNAs from representatives of a net-casting spider and an orb weaver were characterized. Two proteins *MiSp* and *Flag* are exclusively known from Araneoids. But Araneoids and Deinopoids were found to be united by their shared use of the protein *MaSp2*. The analysis of the silk genes supports the single origin of orb webs (Garb et al. 2006) (Fig. 13.3).

13.10 Conservation of Spider Silk Protein Genes

There are 34,000 species of spiders characterized by: (1) Each species uses silk. (2) Some ecribellate orb weavers (Araneoidea) have a varied tool kit of task-specific silks with divergent mechanical properties. (3) Each of the fibers is composed of one or more proteins encoded by the spider's DNA (the silk fibroin genes). (4) The fibers are drawn from dissolved fibroin proteins that are stored in specialized abdominal glands. (5) All the amino acid motifs that are conserved in spider fibroin proteins are found in other structural proteins that also have exceptional mechanical properties. (6) Silk production occurs in insects such as moths but is formed not on the abdomen but in labial glands (Gatesy et al. 2001). (7) The repetitive sequences of fibroins (*MaSp1*, *MaSp2*, *MiSp* and *Flag*) from orb-weaving spiders have been maintained over 125 million years of evolutionary history (Fig. 13.3).

The retention of these conserved units and their appearance in other structural superproteins imply that these amino acid sequences are of central importance in understanding the exceptional properties of orb weaver silks.

13.11 A Spider Has Not a Static Mind: The Evolution of the Web

Specialists who studied spiders for many years were puzzled by the building of different types of webs. The result was a series of contradictory interpretations of their origin and evolution. Some considered the spiral web as a late arrival resulting from experience accumulated during previous building of simpler webs. Others were inclined to think that the spiral appeared first and was later modified.

What is independent of the various interpretations, is the finding that there has been an enormous variation in web techniques and solutions to catch prey. This is a demonstration that both processes were far from static but have evolved. An example is the radial web which has led to many different solutions. In its simpler form it is merely a retreat with threads radiating from the opening of the web (Cribellate spiders like *Filistata* and Ecribellate such as *Segestria*). An expansion of this type of web is found in the three-dimensional forms (e.g. *Linyphia*). In another solution the web becomes centric and the spider now sits in the hub. Another invention is the spiral web.

The way the glue was used to trap insects has also evolved. In the simple cobwebs glue is only found in droplets, but in the spiral webs a second sticky silk spiral is laid upon. The Australian spider *Dicrostichus* uses a short thread with a large drop of glue on its end. Orb webs without any sticky spiral are typical for the American species (*Wixia*). As Foelix (1996) states: "These examples show that evolution has not stopped at, but has gone beyond the orb web" and he asks a most pertinent question: "A very basic issue not dealt with previously is how spider webs per se could have evolved". Silk threads are considered to have been originally a means to detect and not to trap prey, sticky silk appearing later as an improvement.

The silk threads produced in the abdominal glands consist of alpha-keratin crystals embedded in a rubber-like matrix of amino acid chains which are linked to the keratin structures. These dry threads form the spokes of the web. The sticky spiral consists of wet threads that have glue droplets surrounding glycoprotein doughnuts located at intervals along its length.

We tend to think that the ability to predict adverse situations and to protect one's property are solely human attributes, but the spiders do it equally well. They coat the threads of their webs with fungicides and bactericides which prevent microorganisms from consuming webs. The spider's mind is able to predict and counteract adverse situations.

13.12 The Geometry of the Web Follows Structural and Functional Economy

The orb web is built with a minimum of material (0.1–0.5 mg of silk) and time (30–60 min). In addition, the strung radial threads converging in the web's center have two advantages: They transmit vibration signals from the periphery to the center and they provide direct access routes.

Equally remarkable is that tensions in the orb web are not distributed at random, but are highest in the periphery and lower towards the center. The forces acting on the threads have been measured in micronewtons (the standard unit of force). The thickness of the threads has also been adjusted by the spider, the result being that higher tensions are balanced by stronger threads (Foelix 2011).

13.13 Different Species of Primates, Like Different Species of Spiders, Employ Diverse Solutions

The spider *Araneus diadematus* is a common species known for building a most complete spiral web. But no other spider family uses the artifice displayed by *Scytodes* to capture game. When approaching a fly it seems indifferent and stays still, suddenly it spits a net made of gum over the fly paralyzing it—a totally different solution but equally efficient. Yet there are other ways of catching prey. Why use spirals and nets? Instead *Dysdera crocata* moves by night and attacks prey directly with its sharp "teeth".

Primates behave similarly coming to different solutions depending on the species. Humans are obsessed by mathematics and geometry, whereas chimpanzees specialize on making more than 30 distinct calls including the panthoot which consists of shrieks and roars that can be heard 2 km away. Besides, each adult chimpanzee builds a new, individual tree nest each night for sleeping. Gorillas, instead, have a typical nervous behavior. They bark to deter attacks and walk quietly in single file into thick forest to avoid danger (Burnie 2004).

Differences in the minds of spiders and primates result in quite diverse solutions.

13.14 The Similarities Between Spiders and Humans

The similarities are as follows: (1) Both spiders and humans can draw a spiral. (2) The spiral cannot be drawn without previously being conceived by the brain. (3) The spirals drawn by humans are of different types. This is also true of spiders which built different types of silk spirals. (4) The brain of a spider uses the same neurotransmitters that are used in the human brain. (5) Different species of spiders build different types of webs as humans construct different types of tools. (6) Web-building, as well as the use of glue, have evolved. Humans during their evolution have also improved their tools from stone axes to swords. (7) Drug intake in spiders and humans leads to disturbed activity and distorted tool-making. (8) Not less than seven different types of silk proteins are produced by the abdominal glands of spiders which have different elastic properties. The spider's brain uses them selectively for different purposes. Humans, as well, employ different proteins in different products. (9) The proteins found in silk are produced by specific genes. One of the components of silk threads is the protein keratin which is the main protein in human hair and skin (Lackie and Dow 1995).

Sources of Figures

Fig. 13.1 Wells, M. 1983. Cephalopods do it differently. New Scientist 1983: 332–337 (Fig. page 337).

Fig. 13.2 Pope, J. 1986. Do Animals Dream? Michael Joseph, London, UK (Fig. page 78).

Fig. 13.3 Gatesy, J. et al. 2001. Extreme diversity, conservation, and convergence in spider silk fibroin sequences. Science 291: 2603–2605 (Fig. 3, page 2605).

14 Hexagons Formed by Molecules and by the Minds of Wasps and Humans

14.1 Hexagons Were Formed by Simple Molecules, Minerals and Living Organisms Before They Appeared in the Human Mind

The geometric figure with six sides called a hexagon, that we draw on a sheet of paper, existed long before it was formed in our minds and was concretized on a drawing. Before humans and large mammals had populated the planet and before minute protozoa inhabited the Earth, the hexagons had arrived. They were part of the structure of atoms.

A snow crystal, that consists of only two of the simplest atoms: hydrogen and oxygen, has already a hexagonal shape. This is not accidental, all snow crystals so far photographed, and they are over 10,000 have this geometric shape. Besides, when snow crystals ramify, building side branches, they continue to preserve the hexagonal form (Fig. 14.1). Hence the building of hexagons is an inherent property of atoms. No other geometric configurations arise when hydrogen and oxygen combine to form water under atmospheric conditions.

Following the evolution of matter, minerals appeared on the surface of the planet. Those in which the chemical element beryllium occurs, such as emerald gems [$Be_3Al_2(Si_6O_{18})$] and chrysoberyl ($BeAl_2O_4$), show that when beryllium combines with the atoms of aluminium and oxygen the hexagon becomes an obligatory shape (Figs. 14.1).

Most flowers are born in a geometric prison. They cannot escape from an original symmetry that has been preserved throughout millions of years and has remained coupled to other plant traits in a vast number of species. The result has been an easy demarcation of plants into families by botanists. Cruciferae are characterized by flowers with 4 petals and 4 sepals (fourfold symmetry). Those of Rosaceae have usually a fivefold symmetry. Liliaceae (Lily family) in which the sepals and petals are similar to each other build a typical sixfold symmetry. The same is the case in the Daffodil family (Amaryllidaceae) with their 6 tepals producing a hexagonal shape.

As evolution reached the invertebrates and vertebrates, the hexagons appeared again this time on the surface of different organs, such as the eyes of insects, or on the skin of fishes, reptiles and mammals (Figs. 14.1 and 14.2).

This regularity is modified in some species, such as the giraffe, where hexagons may become transformed into pentagons and in the reptile *Varanus giganteus* on which the hexagons become successively modified into circles. The transformation of the hexagons into other geometric figures shows that an underlying geometry can take different forms as is the case in the shape of flowers.

14.2 The Atomic Preservation of the Hexagonal Pattern Occurs Independently of DNA

The question that arises is: What mechanism is responsible for the intact preservation of this pattern from water crystals to the mammalian skin.

We tend to attribute most body traits to the action of DNA. But we are obliged to realize that DNA knows nothing about symmetries. It consists only of four bases attached to a back bone built by sugars linked by phosphorylated groups. DNA cannot produce chemical elements such as iron or magnesium, which must be imported from the outer environment surrounding the cell. Proteins and other molecules which have crucial functions in living organisms, such as hemoglobin, have functions which are not located on the large amino acid sequences that they received from the DNA code but from such simple chemicals as iron. Chlorophyll, one of life's key molecules, derives its function from a magnesium atom which is not present in the DNA structure.

It is these and other chemical elements, such as hydrogen and oxygen, which already have an inherent pattern which can only produce a given geometric figure when they combine.

It is thus expected that it is the atomic configuration that the body molecules present to other molecules, at a given time in plant and animal development, that leads to the

A. Lima-de-Faria, *Molecular Origins of Brain and Body Geometry*,
DOI 10.1007/978-3-319-06056-9_14, © Springer International Publishing Switzerland 2014

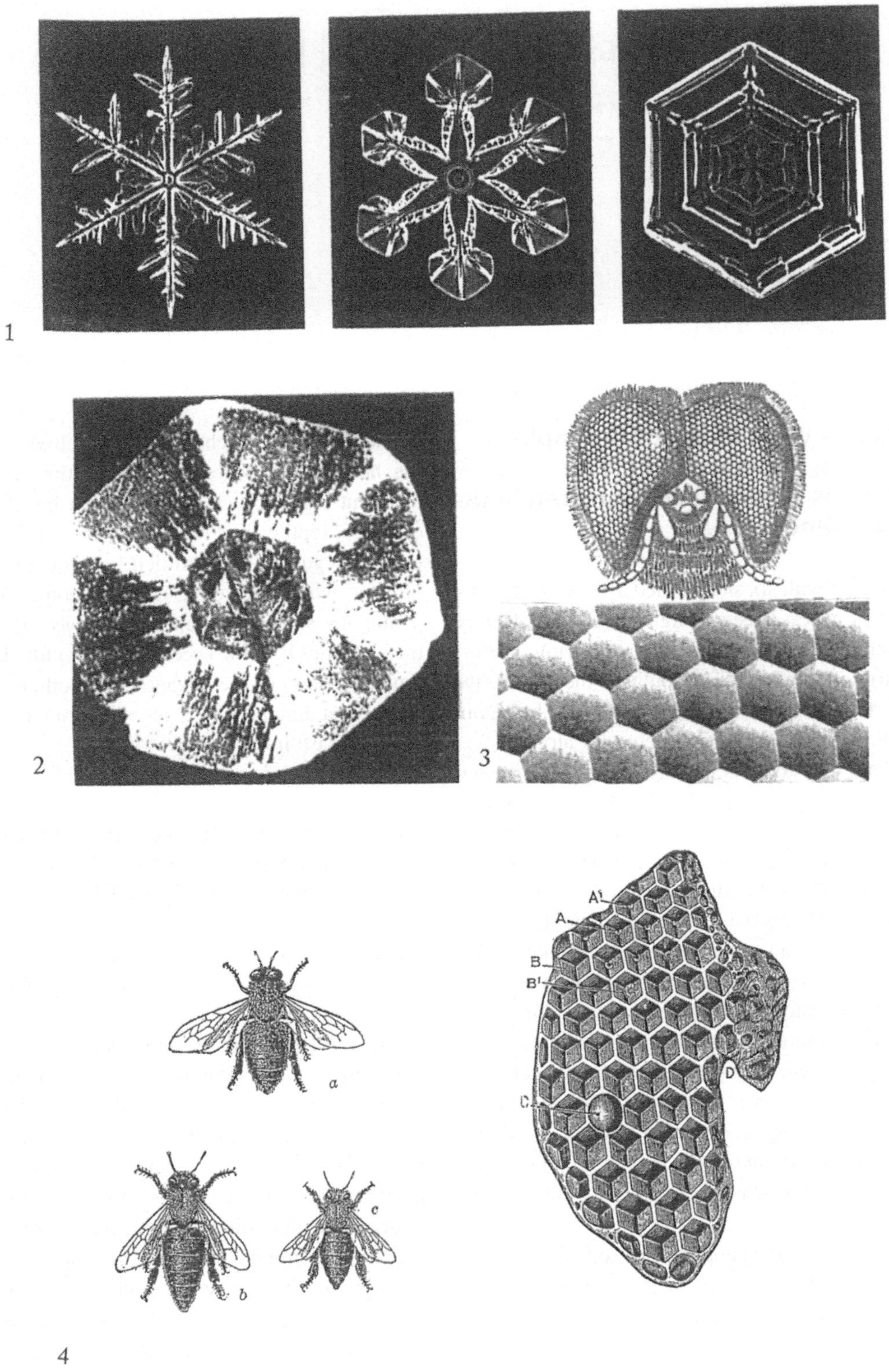

Fig. 14.1 ***Hexagons* are formed by water, minerals and invertebrates.** (**1**) Snow crystals with their characteristic *hexagonal shape*. Snow crystals form ordered ramifications maintaining the six-fold symmetry. (**2**) A crystal of emerald (a beryl gem) with the typical *hexagonal shape* of the beryl mineral. (**3**) The head of an insect and the hexagonal structure of a wasp's eye seen at a higher magnification. (**4**) The common honey bee and the honeycomb it builds

critical decision resulting in a final geometric configuration. As a consequence this is repeated time and again irrespective of the stage of evolution that matter has attained.

Once a given atomic combination repeats itself the same geometric figure emerges in minerals, flowers or animals, whether you like it or not.

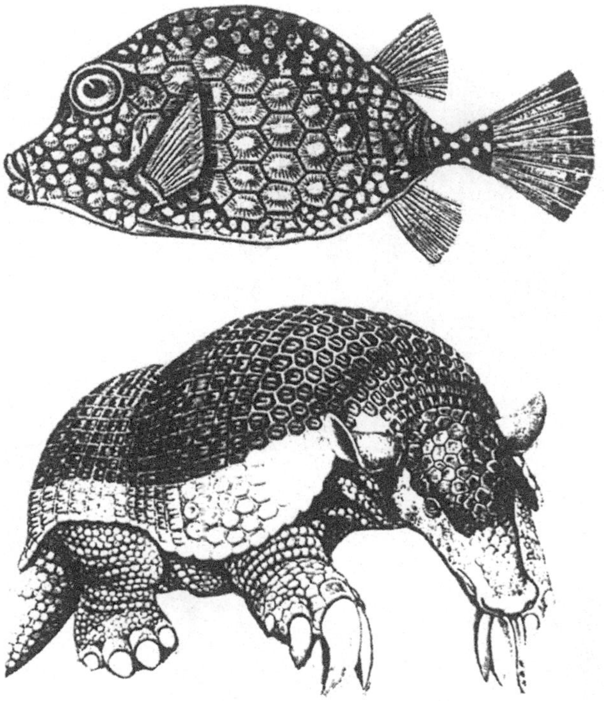

Fig. 14.2 ***Hexagons*** **are formed on the body of fishes and mammals.** (**1**) The fish *Ostracion triqueter*. (**2**) Giant armadillo of South American forests. A mosaic of small bones results in a mantle of *hexagons*

14.3 The Building of Hexagons Has Been Attributed Earlier to Mechanical Causes

Physics in the first decades of the 1900s was such a well-developed science, based on an imposing body of data, that it had a dominating influence on other disciplines such as biology. At that time most phenomena were interpreted on the basis of mechanics and as a result biological phenomena tended to be seen as mechanical events.

Consequently the shaping of hexagons by honeybees (*Apis mellifera*) was attributed to mechanical causes. Various assumptions were based on this idea: (1) The honeybee body, so uniform in size, would provide the measuring system that keeps the small hexagonal cells so regular. (2) Bees built hexagons to economize with wax. (3) Each bee was thought to strive to make a new cell as large as possible and squeeze it against adjacent cells in the construction process. (4) The consistency of the wax was the main source of the hexagonal shape. More than a thousand years before the mechanical interpretation, the mathematician Pappus of Alexandria (AD 320) suggested that the building of hexagons endowed bees "with a certain geometrical forethought" (Line et al. 1983).

14.4 Hexagons in the Mind of Bees and Wasps

At present much more information is available on the behavior of insects. Before, what was described under the name of a "fixed instinct" has been substituted by evidence revealing a versatile behavior that results in novel solutions in face of novel situations. This is not only true of insects but of many other living organisms.

What turns out is that bees may produce hexagons out of wax but under other circumstances their nests may consist of individual circular cells or result in other shapes (Figs. 14.1 and 14.3). It is not the wax that is the obligatory material to build hexagons. Wasps, use instead plant materials that are chewed into a pulp producing hexagons undistinguishable from those of bees in their impressive regularity.

Other species of wasps (*Eumenes coarctatus*) do not care for hexagonal shapes assembled into a nest. Instead they build clay spherical jars having a circular opening. These may be deposited alone on plant branches or be put together in small groups on a board.

The observations on bee and wasp behavior tend to show that we are dealing with an event that occurs in the minds of these insects. The building of a hexagonal shape in their minds has the characteristic of being an atomic event, just as the hexagons that are formed on their eyes. Moreover, as the hexagons may change into circles along the skin of a single animal due to novel chemical arrangements, the mind of a given species may change from conceiving hexagons to conceive circles. There may also be a difference between species as seen in wasps.

A similar behavior arises in vertebrates. Birds weave nests following working sequences that insure stability and circularity. Like a spider, the African gold weaver starts its nest by first building a cross-bridge between two reeds as a support. Grass stalks are then used to make a ring which will become the circular entrance of the nest. This is the same type of solution found in the circular entrance to the nest of the solitary wasp *Eumenes* (Fig. 14.3).

14.5 Nest Building in Birds Leads to Different Mental Decisions: Not All Cuckoo Species Lay Their Eggs on Other Birds' Nests but Some Cuckoos Build Their Own Nests

Since Ancient Greece cuckoos have become part of the folklore due to their parasitic behavior. They do not build nests and instead lay their eggs in those of other birds. These

1

2

Fig. 14.3 (1) **Bees construct nests according to the situation they face.** Cluster of *circular* cells built on the ground by *Lasioglossum malachurum* containing pollen balls and larvae (*left*). Nest of the bee *Osmia aurulenta* which used the shell of a dead snail to build its cells (each with a larva). (2) **African golden weaver, *Ploceus subaureus*, starting its nest, like a spider, by first building a cross-bridge between two reeds as a support.** Grass stalks are then used to make a ring which will become the *circular* entrance of the nest (photographs)

birds are usually much smaller in size and go on feeding the foreign animal that grows fast in their nests and soon becomes ready to fly. This form of behavior has excited the imagination of authors and has been the source of much discussion on altruism in animals. What is seldom or never mentioned is an important observation that has become available as bird studies have proceeded. The parasitic behavior, which demands a mental planning of the egg-laying in a foreign nest, is found within as many as 50 species of the subfamily Cuculinae but is not present within another subfamily of cuckoos, the Coccyzinae which are American cuckoos comprising 18 species (Fig. 14.4).

Fig. 14.4 Two distinct mental decisions occurring in cuckoos. The family Cuculidae consists of several subfamilies. The Cuculinae are parasitic laying their eggs in other species nests whereas the Coccyzinae build nests where they lay their own eggs

Remarkable is that in the subfamily Neomorphinae there suddenly appear 3 species with the same brood parasitism whereas 7 are normal (Ericson and Sjögren 2004). Non-parasitic cuckoos build their own nests, lay their eggs in them and rear their young, as most bird species do, which means that they make a quite different mental decision when spring comes. A third solution is known, in which cuckoos instead of forming solitary pairs, build social groups that breed cooperatively, a situation opposite to brood parasitism. In these social groups some birds are "helpers". They actually help to rear the young of other individuals which represents an altruistic behavior (Payne 2005).

There are birds belonging to other families that do not even build a nest laying their eggs directly on bare soil as does the Standard-winged Nightjar *Macrodipteryx longipennis* (Holyoak 2001).

Four main mental decisions are made by birds that lead to: (1) Making of a nest. (2) No nest-building. (3) Use of a foreign nest. (4) Cooperative breeding.

14.6 Young Cuckoos Migrate Without Knowledge of the Route and Without Parental Guidance: A Mental Imprint Decides Migration

Another surprising mental process has been discovered recently. Once the egg-laying period finishes by the end of July the adult cuckoos immediately migrate southward. They usually arrive in Africa as early as mid-August. In the meantime the young cuckoos growing in the nests of their foster parents become ready to fly. About a month after their true parents have left Europe, the young begin their migration. An unexpected event arises. They do not follow their foster parents in their flights and their true parents are no longer available to teach them. Without knowledge of the route and no parental help, the young cuckoos migrate in a precise direction reaching the African grounds where their true parents arrived earlier (Elphick 2007). This implies the establishment of a mental imprint, but the mechanism that is responsible is not yet known since the phenomenon has only been detected recently.

14.7 Protein and Hormone Changes Occur in the Brain of Singing Birds

Singing demands mental innovation based on order, because any succession of sounds that are melodic exclude randomness—they follow a well-defined mathematical sequence known since antiquity (Chapter 12). This is why listening to the nightingale or a canary results in a sensation of harmony and mental pleasure.

The molecular basis of singing and musical ability is best known in birds and humans. Like human youngsters, songbirds learn to vocalize by imitating the sounds of their elders. When zebra finches learn to sing, or adult canaries change their song, there is an increase in the protein produced by the *FoxP2* gene in the bird's brain. If the amount of this protein is reduced, by using virus-mediated RNA interference, there is a disruption in bird song which becomes more variable and imprecise than that of controls (Haesler et al. 2007).

Other molecules such as hormones participate in the song process (Shen et al. 1995). Neural conversion of androgen to estrogen by the enzyme aromatase is an important step in the development and expression of masculine behavior in birds. The distribution of the aromatase messenger RNA has been mapped in the brain of finches and found to be widely expressed. The presence of estrogen receptors in the brain of several avian species has turned out to be unique to songbirds (Gahr et al. 1993).

Dopamine is a modulatory transmitter in the brain. Its activity is due to the gene *avian zenk* which has been linked to the type of singing in birds (Kubikova et al. 2010).

14.8 The Molecular Basis of Musical Ability and Singing in Humans

Genetic aberrations of the human gene *FoxP2* impair speech production. The protein encoded by this gene is essential for full articulation of the human language. Mutations of *FoxP2* cause verbal disorders. This is actually the same gene that controls song production in birds and whose variation in activity in a bird's brain leads to song disorder (Haesler et al. 2007).

A myriad of studies have been carried out on the heredity of musical aptitude in humans. As a result, several genes involved in human singing and musical talent have been identified. As Stafford (1980) points out several components participate in this process: tonal memory, pitch discrimination, rhythm memory, timbre, fluency and finger dexterity. Hence, it is not one gene but a whole series of genetic markers that are considered to be involved in musical talent and singing. This has led to the extensive investigation of the musical ability in twins to assess the influence of the environment in musical display. It turns out that the environment has a critical and necessary role (Coon and Carey 1989).

More recently molecular techniques were used to unravel the biological background of music perception. Genome screening of 15 Finnish families, searching for microsatellite DNA markers, revealed a major gene locus on the human chromosome 4 (4q22) to be connected with musical aptitude (Pulli et al. 2008). Pitch recognition has also been approached from the genetic point of view and it has been found that there are highly heritable differences in auditory functions (Drayna et al. 2001). Step by step one is coming closer to the location of different genes involved in musical ability.

14.9 The Musical Talent of Whales Compared to That of Humans

Humans are not the only mammals that sing and that can produce melodies. The humpback whale *Megaptera novaeangliae* has been studied extensively using deep-sea hydrophones combined with satellite tracking. These cetaceans migrate to tropical waters during the winter where they reproduce. Male mating strategy involves singing, as is the case in birds, as well as in humans under certain conditions. Lone whale males repeat long complex songs each lasting for about 10 min in an unbroken series. They can sing incessantly for over 24 h. They follow a hierarchical model in which "each song consists of a series of themes repeated in a specific order, the themes in turn are made up of phrases repeated a variable number of times" (Macdonald 2002). Songs from different years are quite distinct from one another. Besides, the songs of different humpback populations are independent. They can even adopt a new song from a different whale locality. Macdonald (2002) concludes that the humpbacks "learn the detailed acoustic structure of their songs". The parallel with human musical ability is striking.

Like whales, birds can also improvise and learn new melodies as did the starling owned by Mozart which was able to whistle one of his concertos (West and King 1990). We face a situation that leads us to the problem of the variation in genetic make up that results in the sudden appearance of individuals with exceptional geometric and musical abilities such as found in highly talented musicians and mathematicians.

14.10 Talented Humans Show an Early Mental Imprint: Music and Geometry Tend to Hang Together

Exceptional persons show mental capacities at an early age and in some cases without the intervention of teaching. It appears as a mental imprint like that found in cuckoos.

Leopold Mozart (1719–1787) was a violinist of high reputation in the service of the Prince of Salzburg (Austria). He composed many symphonies and other orchestral works being also the author of a famous treatise in violin playing called "Violin School" 1756. This work was translated into various languages. Two children were born from his marriage. Anna (born 1751), also called Marianne, was a child prodigy harpsichordist who later became a music teacher in Salzburg (Scholes 1977). Wolfgang Amadeus Mozart (born 1756) showed from earliest childhood such an exceptional talent for music that his father dropped other activities and devoted himself to educating the boy. "He produced his first minuets at the age of five, his first symphony just before his ninth birthday, his first oratorio at eleven, and his first opera at twelve" (Grout and Palisca 2001). Throughout his short life (he died at 35) Wolfgang produced more than 600 compositions.

Another example is the Bach family which is well-known because the musical ability ran uninterrupted for 6 generations, the outstanding member being Johann Sebastian Bach (1685–1750) (Fig. 14.5).

Blaise Pascal was born in Clermont-Ferrand (France) in 1623, his mother dying when he was three years old. The father Etienne Pascal was a distinguished physicist and mathematician who maintained contacts with the leading scientists of the time. Soon it became clear that Blaise had a scientific talent, but his father prohibited him from dealing with science before he mastered Latin—the obligatory language of educated people in the 1600s. However, at the age of 12, he discovered geometry by himself and reinvented the 32 propositions of the first book of Euclid (Flutre 1935).

Pascal's ability to perceive geometry spontaneously is still better described by Crease (2009): "Occasionally, humans have taken the journey on their own, discovering the Pythagorean theorem without the aid of teachers. One is the French mathematician and philosopher Blaise Pascal, whose father forbade any discussion of mathematics around the household, afraid that the subject might distract his child from the all-important studies of Greek and Latin. But the young Pascal began to explore geometry with the aid of a piece of charcoal, in the process discovering many of the proofs codified in Euclid's *Elements*, including the Pythagorean theorem."

Albert Einstein's childhood (1879–1955) was characterized by being late in speaking and writing. From the age of six he began to learn the violin but "seven years passed before he was aroused by Mozart into an awareness of the mathematical structure of music" (Clark 1973). His discovery of geometry occurred about the same time. When 12 he was given a copy of Spieker's book "Lehrbuch der Ebenen Geometrie". After a few months he had worked through the whole treatise. Moreover, "he proved the Pythagorean theorem for himself based on the similarity of triangles" (Crease 2009). Einstein wrote later in an autobiographical essay: "it is marvelous enough that man is capable at all to reach such a degree of certainty and purity in pure thinking".

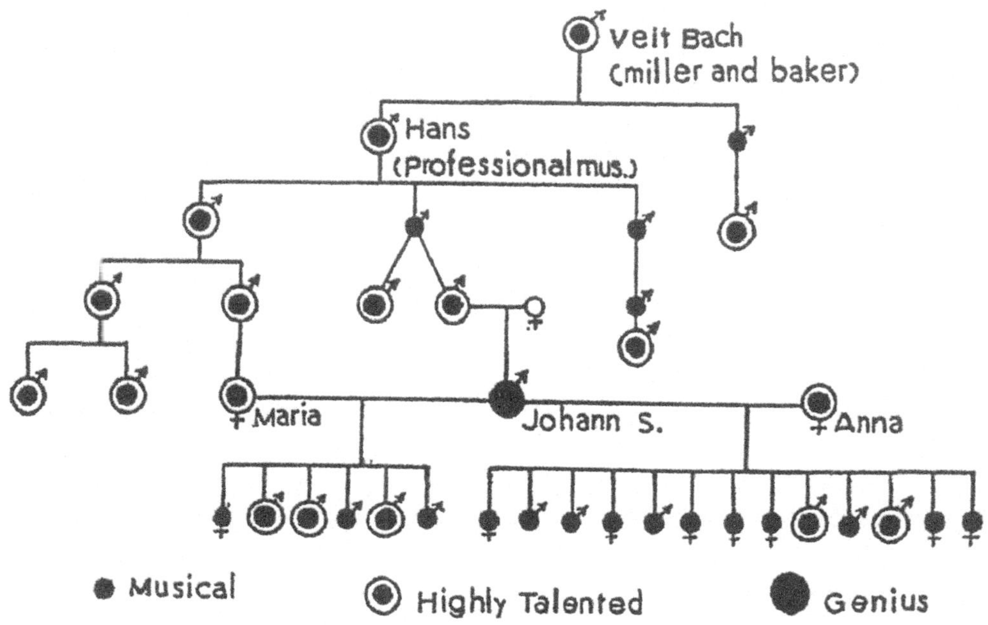

Fig. 14.5 Johann Sebastian Bach (1685–1750) was the outstanding member of six generations of musicians. Vast genetic research of twins, family pedigrees and molecular analysis of DNA, have revealed specific genes which are related to musical performance

Sources of Figures

Fig. 14.1 (**1**) Porzig, W. 1963. Schneekristalle. Zeiss Inf. 47: 28–29. (**2**) Medenbach, O. and Sussieck-Fornefeld, C. 1983. Minerais. Ed. Publica, Lisbon, Portugal (Fig. page 00). (**3**) Björn, L.O. 1973. Ljus och Liv. Aldus/Bonniers, Stockholm, Sweden (Fig. Plate 7). (**4**) Pierantoni, U. 1944. Tratado de Zoologia. Editorial Labor, Barcelona, Spain (Fig. page 657).

Fig. 14.2 (**1**) Pierantoni, U. 1944. Tratado de Zoologia. Editorial Labor, Barcelona, Spain (Fig. 922, page 850). (**2**) Pope, J. 1986. Do Animals Dream? Michael Joseph, London, UK (Fig. page 70).

Fig. 14.3 (**1**) Michener, C.D. 2007. The Bees of the World. The Johns Hopkins University Press, Baltimore, U. S.A. (Fig. Plate 16, page 46). (**2**) del Hoyo, J. et al. (Editors) 2010. Handbook of the Birds of the World, Vol. 15. Lynx Edicions, Barcelona, Spain (Vol 15, Figs. pages 103 and 121).

Fig. 14.4 del Hoyo, J. et al. (Editors) 1997. Handbook of the Birds of the World, Vol. 4. Lynx Edicions, Barcelona, Spain (Vol. 4, Fig. page 509).

Fig. 14.5 Colin, E.C. 1942. Elements of Genetics, 2nd Edition. From Text Material 2011 by Steven M. Carr

http://www.mun.ca/biology/scarr/Bach-Family-Pedigree. html(Fig. page 1 of 1)

Part IV

The Mind Consists of the Same Matter that Has Displayed Order and Coherence Throughout Its Evolution

15 It May Sound Outrageous but the Human Condition Appears to Be Anchored in the Organization of Galaxies

15.1 Astrophysicists Describe Stellar Evolution in a Way Essentially Similar to Biological Evolution

The simplest observation that astronomers make is that stars shine. The sources of this energy and the changes in stellar structure that occur with time are the basis of stellar evolution. As Sears (1981) states: "The theory of stellar evolution is sufficiently developed that each of the stars we can see can be identified with a particular evolutionary stage".

It turns out that stellar evolution has many features in common with biological evolution. Even its representation takes the form of phylogenetic trees like those used for living organisms (Fig. 15.1).

The three principal energy sources in stars, in order of occurrence during a typical star's life, are: (1) gravitational energy (due to attraction between bodies), (2) thermonuclear reactions (proton-proton interactions) and (3) thermal energy (particle movement). Each of these has been estimated for the Sun as well as for other stars. The chemical composition of the Sun has also been determined to be about 94 % hydrogen, 6 % helium, and 0.1 % of all the heavier elements.

The Sun's gravitational contraction can supply it with energy to continue to shine for 8×10^9 years, but nuclear reactions provide the principal energy source consisting of hydrogen burning into helium and helium burning into heavier elements.

The age of the Sun is also known. The formation of our solar system occurred 4.5 billion years ago and the universe started almost 14 billion years ago (Scharf 2012). Hence, there is a time scale for stellar evolution as there is one for biological evolution. The difference is mainly one of magnitude.

15.2 The Graphic Representation of Stellar Evolution

Stars emit radiation over a wide range of wavelengths (stellar spectra) which allows them to be divided into spectral classes blue-white, white, white-yellow, yellow, orange and red. The Sun is a yellow type star.

In the blue region of the spectrum are *hot stars.* Occurring at the red end of the visible spectrum, or in the infrared, are *cool stars.* Stellar spectra indicate temperature, as well as chemical composition and star mass, because the mass and temperature of a star are related to its chemical composition.

Stars were created with a high temperature but cooled as they passed through four stages: (1) *Bluish stars* (ex. Sirius), (2) *Yellow stars* like the Sun, (3) *Red stars*, often variable (ex. Betelgueuse), (4) *Redder* than the third type and of low brightness (not numerous).

E. Hertzsprung and H.N. Russell, working independently, arrived at the same basic description of the evolution of stars. The result of their work is depicted in the Hertzsprung-Russel diagram based on the discovery of a relationship between the density of stars and their luminosity.

The diagram shows the variation of luminosity in stars against spectral class. The great difference in brightness between *giant* and *main sequence* stars derives from a difference in density. The *youngest stars* (blue ones) lie in the upper end of the diagram. *Blue stars* are not only the youngest but the most massive (Scharf 2012).

The data from stellar evolution led to the conclusion that stars left the *main sequence* to become *red giants*, shrinking again and finally turning into *white dwarfs*. These, located at the lower left of the diagram, are degenerate stars of low

A. Lima-de-Faria, *Molecular Origins of Brain and Body Geometry*,
DOI 10.1007/978-3-319-06056-9_15, © Springer International Publishing Switzerland 2014

luminosity which have exhausted all possible fuel used in thermonuclear fusion (Fig. 15.1).

As a star burns its hydrogen core, its radius increases. In due course they become *pulsars* and *black holes*.

Stellar evolution is on a time scale in the order of a trillion years: the time of a star to become a *white dwarf* (North 2008).

15.3 Dark Matter Occupies a Larger Part of the Universe than Light

The past 10 years have witnessed a change in the way astrophysicists think about massive black holes. They are now considered to have a major role in the evolution of galaxies. Although we tend to consider the universe as consisting mainly of light emanating from stars, most of the universe, about 90 %, is considered to consist of dark matter, i.e. matter that does not emit light. Massive black holes may have masses up to billions of times that of the Sun.

Stars and gas represent the content of galaxies in contrast to *dark matter* that does not interact electromagnetically but only gravitationally with its environment. The mass content of the universe is dominated by cold dark matter. Massive black holes are the remnants of the first generation of stars, formed out of pristine gas, which did not yet contain heavy elements.

Black holes reside in the center of galaxies including the Milky Way. Actually their emergence can be dated and their evolution followed through cosmic history. They were formed 13 billion years ago and grew within their host galaxies during this period of time (Volonteri 2012).

15.4 There Is Order in the Milky Way

Nicolaus Copernicus (1473–1543) disposed of the prevailing idea dominating the Middle Ages that the Earth was the center of the universe, by showing that it rotated around the Sun.

Subsequently it was thought that the Sun occupied a position near the center of the universe. By the 1900s it became clear that our solar system was not at the center of the galaxy *Milky Way*. There seems to be no center for the universe because this galaxy is merely one of the many that populate the cosmos.

The center of the cosmos was shifted from the Earth, to the Sun, to the Milky Way and finally to nowhere. This is an example of how ideas change in science as the technology available allows us to penetrate deeper into the realm of matter.

Significant is that it soon turned out that there was order in the Milky Way. In the Orion region, of the great spiral disk of this galaxy, stellar systems are spaced with an average of about 5–10 light-years between them. The central region of this monumental cosmic object is an elongated structure which contains a high density of old red and yellow stars.

Additionally, since nothing in the universe is static, there is a constant movement of stars inside the galaxy. It seems like chaos but, their orbits are only around certain centers. One star circles through its closest approach at velocities of 7,500 miles a second. By comparison the Earth orbits the Sun at less than 20 miles a second.

This picture of the ordered organization of galaxies is not an accidental event. It is the result of the fast development in the study of the skies which has permitted predictions concerning the composition of stars. An example is *neutron stars* which were conceived before they were identified. The formation of neutrons would lead to a star of a very small radius and of extremely high density. The proof emerged when *neutron stars* were discovered. *Pulsars* are examples of such stars (Pitt 1988; North 2008).

15.5 "As a Species We Are Born Out of 4 Billion Years of Fierce Molecular Evolution"

This statement made by the astrophysicist Scharf (2012) is the conclusion that he reaches after treating at length the origin and evolution of stars.

He actually describes how the evolution that led to the human species is anchored in the primeval evolution of matter. The process is not a chaotic one as it turns out that matter in the cosmos is ordered already in its original state and that this order is critical in shaping the future evolution of matter all the way to the most complex biological levels.

The original source of this process is considered to take place in what astronomers call *black holes*. As Scharf (2012) states: "Black holes play a critical role in making the universe appear the way it does. Because of this, they also profoundly influence the environments and circumstances in which planets and planetary systems are formed, and the elemental and chemical mixes that go into them. Life, the phenomenon of which we are a part, is fundamentally connected to all these chains of events. Saying that black holes have implications for life in the universe may sound outrageous and far-fetched, but it appears to be the simple truth."

1

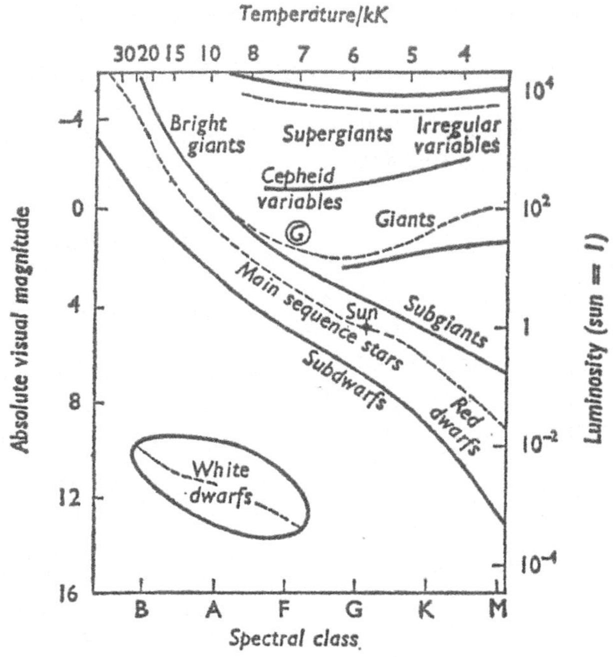

2

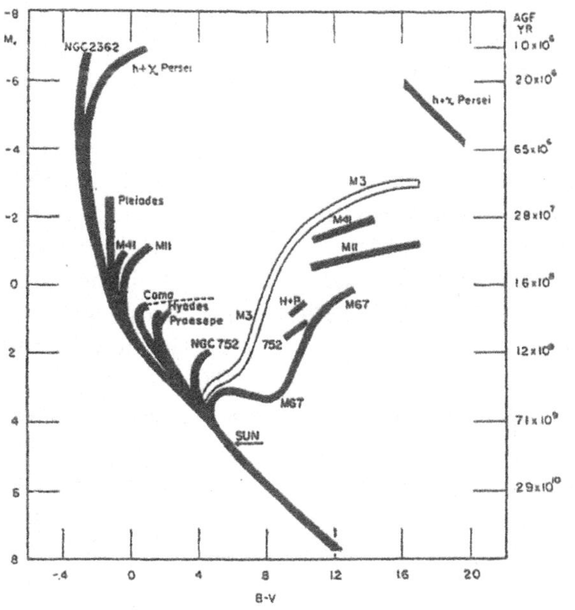

Fig. 15.1 The evolution of stars and galaxies. (**1**) The Hertzsprung–Russel diagram showing the variation of brightness in stars (luminosity) against spectral class. The temperature of a star is related to its chemical composition. The *youngest stars* lie in the *upper end* of the diagram. The branches where several types of giant stars are located contain clusters of *old stars*. As a star burns its hydrogen core, converting it into helium, its radius increases becoming a *giant star*. The different spectral classes are: **B** (*blue–white*), **A** (*white*), **F** (*white–yellow*), **G** (*yellow*), **K** (*orange*) and **M** (*red*). As seen in the diagram the Sun belongs to the **G** class. Most stars lie along a well defined band called the *main sequence*. During stellar evolution the stars *left main sequence* to become *giant stars* and finally turned into *white dwarfs*. *Red stars* come in two kinds: *luminous giants* and *dim dwarfs*. (**2**) The evolutionary behavior of stars in several open clusters of different ages. *Absolute magnitude* (and also *estimated age*) is on the *vertical scale* and color index **B-V** (a measure corresponding roughly to spectral type) on the *bottom scale*. The common *main sequence* is the stem of the tree. Each branch belongs to a different star cluster. Departure from the main sequence is considered to be the result of their evolution. Most branches have missing parts like in the phylogenetic trees that describe biological evolution. *NGC2362*, *M3*, *M41*, and so on, are the names given by astronomers to supernovas (exploding stars) and other star clusters that include galaxies. The *Pleiades* group is a typical open or galactic star cluster (located on the left branch of the tree). Photographs taken with large telescopes show that there is a beautiful gas-cloud spread between its stars

Moreover, it turns out that co-evolution is not a term that belongs solely to the domain of biology. Astronomers, studying the evolution of the universe were confronted with a similar phenomenon. They found that giant black holes and their host galaxies have co-evolved (Scharf 2012).

15.6 No Structure or Function Comes from Nowhere

Nothing seems to come from nowhere.

Once a structure or a function are recognized, it turns out that they existed at a previous level of organization in the same or a simpler form. As one probes into earlier levels, it is soon found that the same properties are present as early as the dawn of the emergence of matter.

The bilateral symmetry displayed by the body of most vertebrates is so rigid and so general that, in encyclopedias, species are depicted in profile. It is taken for granted that the other half of the body is identical to that represented in the figure.

It turns out that this symmetry, that is an integral part of geometry, does not arise from a chaotic vacuum, but from previous patterns that are inherent to matter. The veracity of this statement is exemplified by the right-handed and left-handed forms, which are patent in humans but can be traced all the way back to the galaxies.

15.7 Right-Handed and Left-Handed Forms Are Found from Primeval Matter to Higher Organisms Including Humans: The Original Order Could Not Be Erased

Galaxies have colossal dimensions yet they are organized structures that possess a nucleus accompanied by long arms, that tend to be symmetrical, often building a spiral. Galaxies are not static either, they grow by accreting gas from their surroundings and by converting gas into stars at high speed (Lehner and Howk 2011).

Astronomers and astrophysicists have, for decades, described and photographed two galaxies which are typical of the right-handed and left-handed forms. These are *M.101* (right-handed) and *M.74* (left-handed). In them, convulsive atomic reactions lead to the birth of new stars, yet they maintain the spiral shape that is the hallmark of their identification (Fig. 15.2).

At first, the shape that the galaxies assumed seemed to come from nowhere. However, the study of the elementary particles, that are at their origin, revealed that neutrinos are right-handed and left-handed and that their spin of rotation can be clockwise or counter-clockwise (Pagels 1982). As von Baeyer (1986) put it "The study of matter has become a natural history. Physicists have discovered that there is a story to matter, a story in the true sense of the word".

Since atoms result from the combination of elementary particles, the carbon atom itself turns out to be both right-handed and left-handed, as is the case in glyceraldehyde.

At the next level of atomic organization the crystals were formed. The French chemist and microbiologist Louis Pasteur (1822–1895) is best known for his impressive contribution to medicine. Less known is that, in the early days of his scientific career, he investigated the fermentation of wine, a process that was not understood at the time. The crystals of tartrates (compounds of tartaric acid) turned out to occur as prisms. Pasteur soon noticed that all tartrates occurred in a right-handed and a left-handed form. In addition, when dissolved in water, and submitted to polarized light, the first type of crystals rotated the plane of light to the right and the second type to the left. What perplexed Pasteur was that the mould *Penicillium glaucum* was able to distinguish the minute difference between the two types of crystals and only fermented the right-handed ones (Blaringhem 1923).

Atoms from many other compounds, following their self-organization, have the capacity to produce these two types of mirror images. Although they are chemically identical, mirror image pairs frequently have different biological effects. When they interact with liquid crystals one form might make the crystal violet becoming a spiral in one direction, while the other form will turn it red and wind in the opposite sense (Feringa 2002; van Delden and Feringa 2002).

As the macromolecules were formed the amino acids, which are the atomic building blocks of proteins, appeared in left and right shapes. Cellular processes have also been able to distinguish between the two mirror images. The 23 amino acids which are components of proteins only belong to the left form (Sharp 1988) (Fig. 15.2).

This atomic property did not spare DNA. In cells, this nucleic acid occurs mainly in the right-handed form, but at high salt concentrations DNA assumes a zigzag look becoming a left-handed helix (Hill and Stollar 1983) (Fig. 15.2).

From the cell one arrives at whole organs and whole organisms. In different plant species, cross sections of buds or stems, show the leaves to be arranged in a right or in a left sequence (Fig. 15.2). As one reaches the invertebrates, the snail *Lymnaea peregra* has been the example of choice for many years. This phenomenon occurs so early in the development of the organism that it can be recognized by the orientation of the chromosomes in the mitotic spindle at the second division of the egg (Fig. 15.2). This orientation is actually controlled by a single pair of genes, the type of coiling of the snail's shell depending on the genetic set-up of the mother (Sturtevant 1923; Gilbert 2000).

Last we arrive at vertebrates and at humans. One egg twins are usually similar in most morphological and functional characters, but they may differ in the symmetry of their hair distribution and in the way they use their hands. The Swedish identical twins Gerd and Monica were mirror

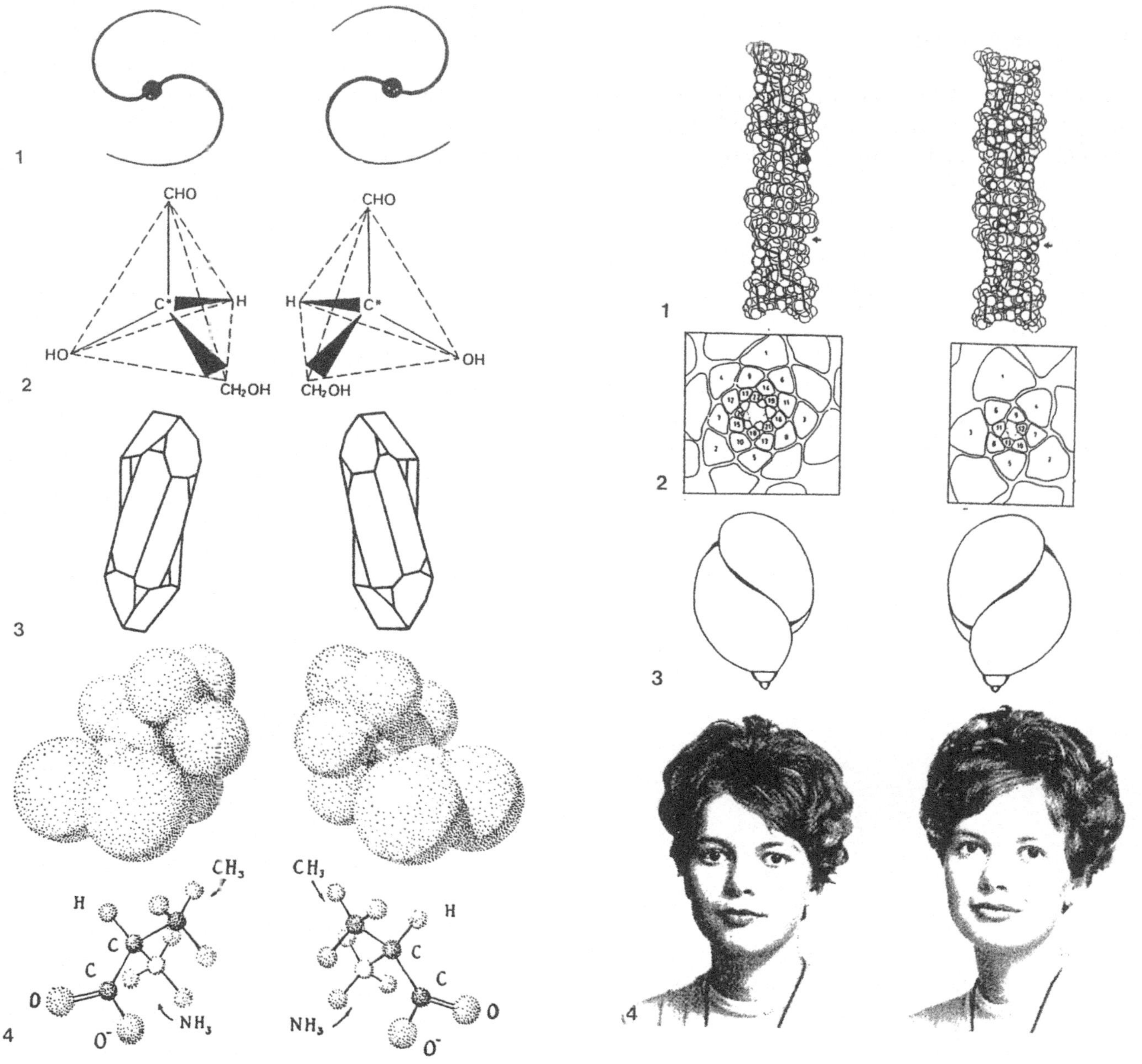

Fig. 15.2 **Occurrence of left-handed and right-handed structures from galaxies to humans.** *Left side*. (**1**) Spiral galaxies of the S-type (left-handed) and Z-type (right-handed). (**2**) Carbon atom bonds in left-handed and right-handed glyceraldehyde. (**3**) The mineral *quartz* in the *left* and *right* forms. (**4**) The amino acid alanine in the *left* and *right* form. *Right side*. (**1**) DNA (Z configuration) in its *left* (unmethylated) and *right* (methylated) forms. (**2**) Sections through plant shoots (conifer *Araucaria excelsa*) showing spiral divergence left- and right-handed. (**3**) Shells of the snail *Limnaea* showing left and right orientation. (**4**) The identical twins Monica and Gerd. Monica is left-handed and has the forelock to the right. Gerd is right-handed and has the forelock to the *left*. They are mirror images of each other

images of each other. Gerd was right-handed and had the forelock to the left and Monica was left-handed and had the forelock to the right (Müntzing 1961) (Fig. 15.2).

The continuity of this phenomenon from elementary particles to atoms, to crystals, to macromolecules and to living organisms, implies the existence of an autoevolutionary process that can not be erased and is sustained as far as the human species (Lima-de-Faria 1988, 1995).

What stands out from the evolution of matter is that, at the same time that a high degree of variation has taken place, a tremendous permanence has obliged a given shape to remain unaltered. Rigidity combined with plasticity is not incompatible. On the contrary they seem to be the obligatory sides of the same coin. Variation can only occur within a well-defined frame otherwise the structure, or function, would collapse.

Sources of Figures

Fig. 15.1 (**1**) Pitt, V.H. (Editor) 1988. The Penguin Dictionary of Physics. Penguin Books, London, UK (Fig. page 180). (**2**) North, J. 2008. Cosmos. An Illustrated History of Astronomy and Cosmology. University Chicago Press, Chicago, U.S.A. (Fig. 231, page 618).

Fig. 15.2 Lima-de-Faria, A. 1995. Biological Periodicity. Its Molecular Mechanism and Evolutionary Implications. JAI Press Inc. Greenwich, Connecticut, U.S.A. (Figs. 1 and 2 pages 122, 123).

16 The Evolution of Matter Is Characterized by Geometry and Permanence

16.1 Only Four Fundamental Forces Direct the Construction and Function of Matter: Symmetries Are Present from the Onset

Discussing the latest results on elementary particle physics, in connection with the discovery of the *Higgs boson*, the astrophysicist Livio (2012) remarked that "Symmetries lie at the heart of the laws of nature", adding that symmetries are found explicitly at the basis of modern physics from general relativity to quantum field theory.

Of importance is that the identification of the new particle also helps to elucidate the way the four fundamental forces, that are recognized in the universe, separated from each other. The weak force is responsible for the release of particles in radioactivity. The strong force glues neutrons and protons together into atomic nuclei. The electromagnetic force binds electrons into atoms and drives chemical reactions. Gravity is the most elusive of the four because it pervades all matter (Glashow 1997). The boson appears to have been instrumental in breaking the symmetry between the electromagnetic and weak nuclear forces.

The existence of just four forces is quite contrary to a system of disorder in which the universe instead would have consisted of a long array of forces with unspecified functions. Hence, the order is patent from the very onset.

We tended to expect random events where there had not been enough research to elucidate how matter behaves. As research became more refined where randomness had been considered to reign order became evident. Hey and Walters (2003) show how powder particles sprinkled over a kettledrum are not randomly distributed, when the instrument vibrates, but build a specific pattern for each type of vibration. The powder collects near the "nodes" where the vibration is weakest (Fig. 16.1). As they also point out, these patterns are analogous to the quantum probability patterns found in electrons. Thus, the ordered way elementary particles behave is transmitted to higher levels of organization in a palpable form.

16.2 The Origin of Permanence: Quarks and Leptons Build a Periodic Table that Antecedes that of the Chemical Elements

A question that arises is: from where does the permanence in the organization of matter derive? The evidence points to the primeval stages of the sub-elementary particles.

Several generations of physicists have worked at the European Nuclear Research Center (CERN) in Switzerland in an attempt to unravel the inner structure of protons and neutrons. Following decades of exhaustive research they had discovered several hundred particles. Confusion reigned, no one could make sense of this "proliferating zoo" as it was then called.

Having in mind Mendeleev's pioneer arrangement of the chemical elements into rows and columns in the Periodic Table, that had led to the prediction of the chemical elements *gallium* and *germanium*, a similar approach was implemented to look for order among a chaotic assembly. The result was that the initial plethora of particles was reduced to two fundamental ones: quarks and leptons, which by combination produced all the others.

The result was the establishment of a modern Periodic Table of Elementary Particles. As Padamsee (2003) points out: "Once again symmetry showed the way to an underpinning structure". Murray Gell-Mann (1929–   ) succeeded in elucidating from the chaotic subnuclear landscape certain underlying patterns well hidden among the new properties and he adds: "Order does exist, although at a deeper level than immediately recognizable" (Fig. 16.2).

A. Lima-de-Faria, *Molecular Origins of Brain and Body Geometry*,
DOI 10.1007/978-3-319-06056-9_16, © Springer International Publishing Switzerland 2014

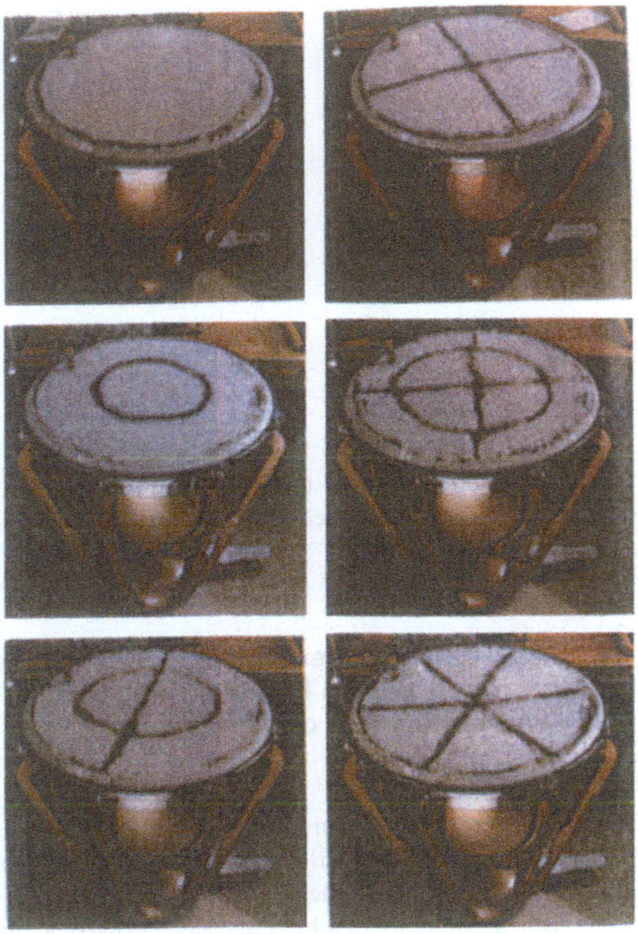

Fig. 16.1 Visible geometric patterns analogous to those of electrons. "A kettledrum sprinkled with powder reveals six of its many vibrational patterns. The powder collects near the "nodes" where the vibration is weakest. These patterns are analogous to the quantum probability patterns for electrons in a box" (Hey and Walters 2003, p. 59)

16.3 Atoms Which Were Considered Imaginary Became Photographed with Powerful Electron Microscopes

Matter continued to evolve by combination. The nucleus of an atom was built by the quarks assembling into protons and neutrons. A proton is composed of two *up* and one *down* quark, whereas a neutron consists of one *up* and two *down* quarks. The next step was the addition of leptons to the nucleus, such as the electrons, leading to the formation of chemical elements like oxygen.

In the early days of physics, atoms were considered imaginary structures that could not be visualized. With the improvement of electron microscopes atoms became palpable and can be seen, one by one, on electronmicrographs. At this level of organization they are packed in regular geometrical patterns.

16.4 The Location and Movement of Electrons in the Atom Follow Rigid Order

Another area plagued by confusion was the distribution of the many electrons that circle around the nucleus of an atom. The negative electrons were assumed to be spread around the central positive atom nucleus. According to classic physics this system would be unstable and would disintegrate. Niels Bohr (1885–1962) found, following his work on the spectroscopy of atoms, that electrons could not move in all directions and could not occupy positions between the orbits that they described around the nucleus. Besides, electrons could only adopt multiples of a certain energy value. These sharp constraints were found to be part of every chemical element and had a key role in any future atomic transformation and combination.

Order in chemical elements took over 100 years to establish. Until 1817 randomness was thought to characterize this area of chemistry. When order started to be perceived elements were placed in groups of three, and afterwards in groups of seven. Subsequently, the similarities of properties were so evident as to allow predictions to be made concerning missing elements. It was Mendeleev who, in 1869, established a relationship between the atomic weights of the elements and their physical and chemical properties. In fact he predicted the existence of six elements that had not yet been discovered. Three were soon found while the other three were discovered later. They had the properties predicted by Mendeleev. Soon it turned out that atoms, with quite different atomic weights had the same properties building periods. The length of these periods followed a series of whole numbers: 2, 8, 18, 32 and so on, indicating another source of order which was found to be related to the maximum number of electrons in each shell. It was not until 1922 that Bohr interpreted the Periodic Table in terms of the electronic structure of the atoms (Pauling 1949). It is the electrons of the outer shell that determine the basic properties (Masterton and Hurley 1993).

16.5 The Evolution of the Chemical Elements from Hydrogen Is Now Firmly Established: The Periodic Table

The evolution of matter received its strongest confirmation when it was established that the over 102 known elements were found to be all derived from the simplest atom hydrogen. The Periodic Table is usually depicted by aligning the chemical elements in rows, but it has been depicted in over 700 different ways (Mazurs 1974). One of them, which best puts in evidence the order inherent to

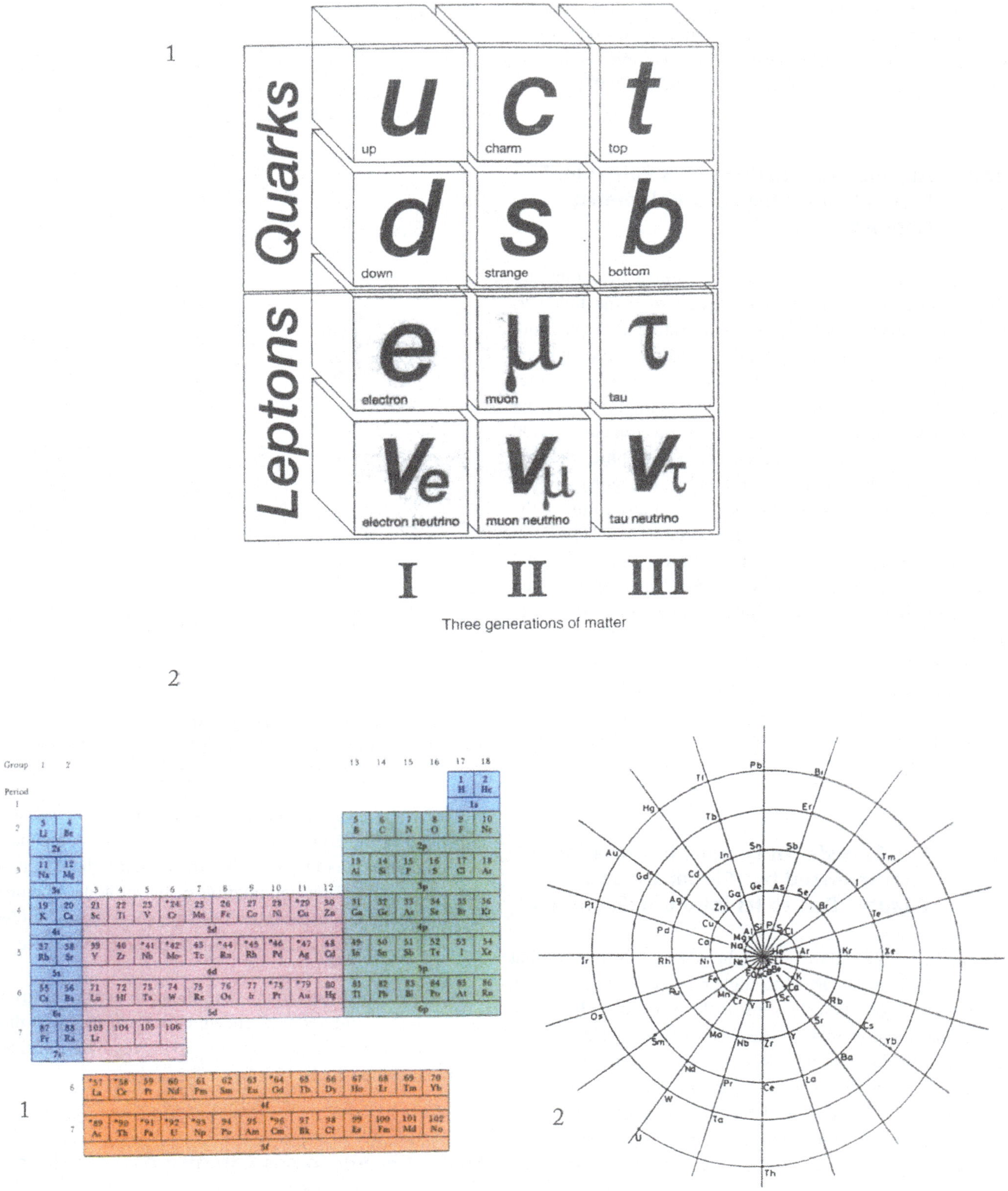

Fig. 16.2 Order at the elementary particle level and at the atom level. (**1**) Periodic table of quarks and leptons showing the production of particles with increasing mass. (**2**) (1) The conventional representation of the periodic table aligning the chemical elements in *horizontal rows*. The table can be used to deduce the electron configurations of atoms. However, there are already exceptions at this level. (2) The representation of the periodic system as a spiral after Erdmann

chemical evolution, is the display of the table as a spiral. Starting with hydrogen, at its center, this atom is soon followed by helium (He), and the other elements, finishing with uranium (U) (Fig. 16.2).

Chemical elements are still being formed at present. Helium and carbon are known to be synthesized from hydrogen in the interior of stars as well as more complex elements: nitrogen, oxygen and the heavier ones (Scerri 2007).

16.6 Atomic Order in Crystals Is Palpable: The Mineral World Exposes Volumetric Geometry

Next came the minerals which were formed by combining the chemical elements. When allowed to come together, under favorable conditions of temperature and pressure, atoms built crystals. Everything around us seems amorph and irregular, but when confronted with a crystal, order cannot be denied. Atoms assemble into configurations from which they cannot depart. As a consequence, the over 3,000 minerals, have built only seven crystallographic systems: cubic, trigonal, hexagonal, orthorhombic, tetragonal, monoclinic and triclinic. The cube is the simplest form of the regular solids on which the other systems are based. No other alternatives were allowed. As expected, these crystal shapes resurface later in the geometry displayed by living organisms since they consist of the same atoms present in minerals (Fig. 16.3).

One should not forget that DNA, RNA and proteins have a crystal structure. It is their ability to build crystals that permitted to decipher (by using X-rays) the way their atoms were distributed within these large macromolecules (Saenger 1988).

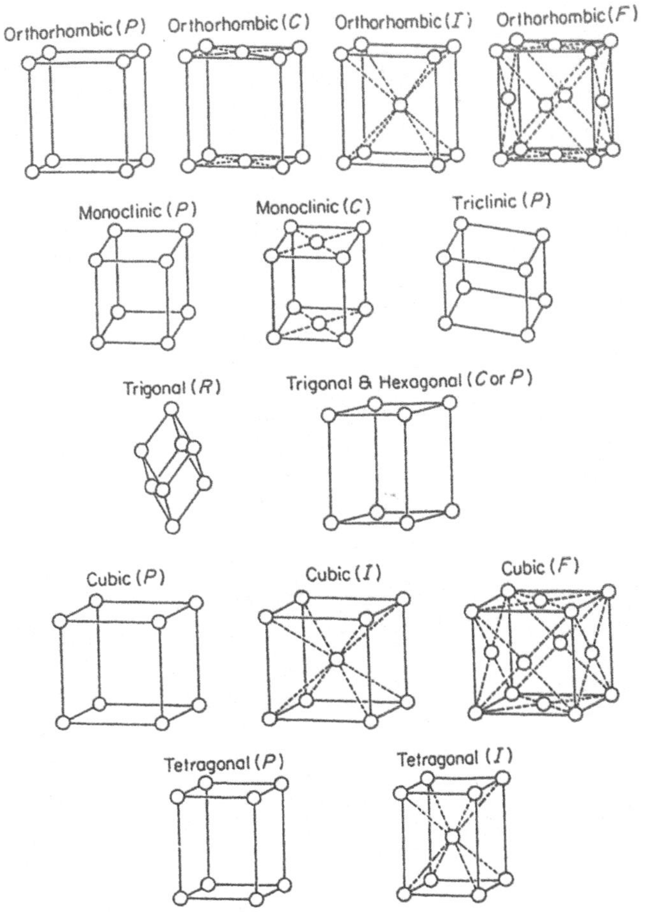

Fig. 16.3 Classification of crystal systems. The 14 unique three-dimensional unit cells. *Balls* represent lattice points around which, or on which, atoms or molecules can be arranged

16.7 The Transfer of Genetic Information Is Characterized by "Simplicity and Beauty" Due to Its Inherent Geometry

The ability of the cell to transfer the genetic information, that is embodied in its DNA, to higher levels of organization resides in the genetic code. This is the mechanism by which the information present in the DNA bases is subsequently transcribed into messenger RNA and later is translated into the amino acids that assemble forming proteins. As Stryer (1981) writes: "The code which is the same in all organisms, is beautiful in its simplicity. Three bases, called a codon, specify an amino acid." Codons are read sequentially by transfer RNAs which serve as adaptors in protein synthesis. The four bases in DNA are: cytosine (C), thymine (T), adenine (A) and guanine (G). RNA contains uracil (U) instead of thymine. The combination of the bases in the transcribed RNA leads to the building of codons: 61 triplets correspond to particular amino acids, whereas 3 code for chain termination.

The occurrence of initiation and termination signals, imbedded in the code's mechanism, is another component that maintains order in genetic information. It is these signals that lead to the exact delimitation of the protein structure. The simplicity and order inherent to the genetic code allows it to be displayed in the form of three concentric circles (Fig. 16.4).

A further step leads to the consideration of the construction of amino acids as independent entities. It turns out that they can be arranged in a spiral due to the regularity of their chemical composition.

16.8 Chromosomes Can Also Assume Near Crystal Form

As one reaches the chromosome, the arrangement of genes and heterochromatin along this cell organelle turned out to be non-randomly distributed. Most genes tend to occur in chromosomes in only a few copies which has made it difficult to localize them with precision on a large number of

The genetic code

First position (5′ end)	Second position				Third position (3′ end)
	U	C	A	G	
U	Phe	Ser	Tyr	Cys	U
	Phe	Ser	Tyr	Cys	C
	Leu	Ser	Stop	Stop	A
	Leu	Ser	Stop	Trp	G
C	Leu	Pro	His	Arg	U
	Leu	Pro	His	Arg	C
	Leu	Pro	Gln	Arg	A
	Leu	Pro	Gln	Arg	G
A	Ile	Thr	Asn	Ser	U
	Ile	Thr	Asn	Ser	C
	Ile	Thr	Lys	Arg	A
	Met	Thr	Lys	Arg	G
G	Val	Ala	Asp	Gly	U
	Val	Ala	Asp	Gly	C
	Val	Ala	Glu	Gly	A
	Val	Ala	Glu	Gly	G

1

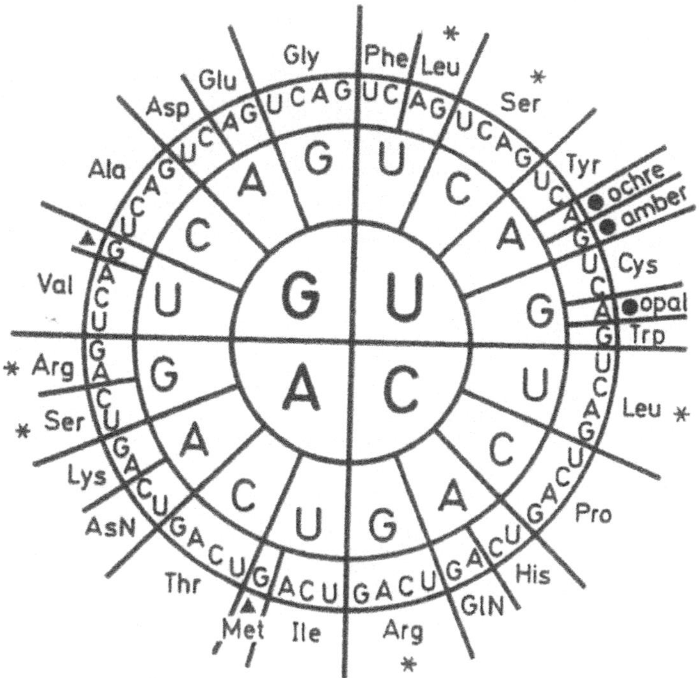

2

Fig. 16.4 The genetic code. The chromosome's DNA is transcribed into messenger RNA. Subsequently the sequence of its bases is translated into a protein sequence, with one nucleotide triplet coding for one amino acid, according to the rules of the genetic code. (**1**) The conventional representation of the genetic code. The four bases present in RNA: *U* (uracil), *C* (cytosine), *A* (adenine), *G* (guanine). The genetic code is characterized by a direct relationship between the sequence of bases in the DNA (and RNA) and the order of amino acids in the polypeptide synthesized from it. Given the position of the bases in a codon (a triplet of bases) it is possible to find the corresponding amino acid (abbreviated as: *phe*, phenylalanine and others). (**2**) The genetic code displayed in radial form. The codons are to be read from the center and represent messenger RNA triplets coding for amino acids entered at the periphery. Amino acids marked with an *asterisk* appear twice, *small solid dots* and *solid triangles* indicate terminator and starting codons, respectively

animal and plant species. For this reason, cytologists have tended to consider them as randomly distributed along chromosomes. This is not the case with the genes responsible for the formation of ribosomal RNA which have been shown, by molecular studies, to be situated at the site where the conspicuous nucleolus is attached to the chromosomes. The analysis of over 506 species of plants and animals disclosed that these genes tend to occupy a position in the chromosome which is close to the ends (telomeres). As the chromosome arms increase in length, the genes preserve this close association with the telomeres. The distribution of the ribosomal genes is so regular that their location can be predicted by means of an equation (Lima-de-Faria 1973) (Fig. 16.5).

The distribution of the large chromomeres that are part of the heterochromatin, located on both sides of the centromere, can be depicted as a spiral due to its regular distribution (Fig. 16.5). Other DNA sequences show ordered distribution along chromosomes.

Like minerals, chromosomes can assume the form of crystals. This is not so surprising since their main components, DNA and proteins, also crystallize. When chromosomes are obliged to be packed into smaller compartments such as spermatids, they form first regular patterns and later assume a near crystal structure as revealed by electron microscopy.

16.9 Permanence in Gene Order: Some Genes Remained Unaltered for over 1.6 Billion Years

Permanence and order in chromosomes has taken additional solutions. An appreciable number of genes, have maintained their DNA sequences practically unaltered since the first cells appeared 1,600 million years ago. The genes for ribosomal RNA are to be found in humans as well as in bacteria and in all other organisms. Another example is the homeotic genes which are responsible for the segmentation of body parts. They occur in plants, invertebrates and vertebrates including humans. Like the ribosomal genes, they have varied in chromosome location, but the integrity of the gene can be easily recognized by its main base sequences and its function (McGinnis and Kuziora 1994). These could

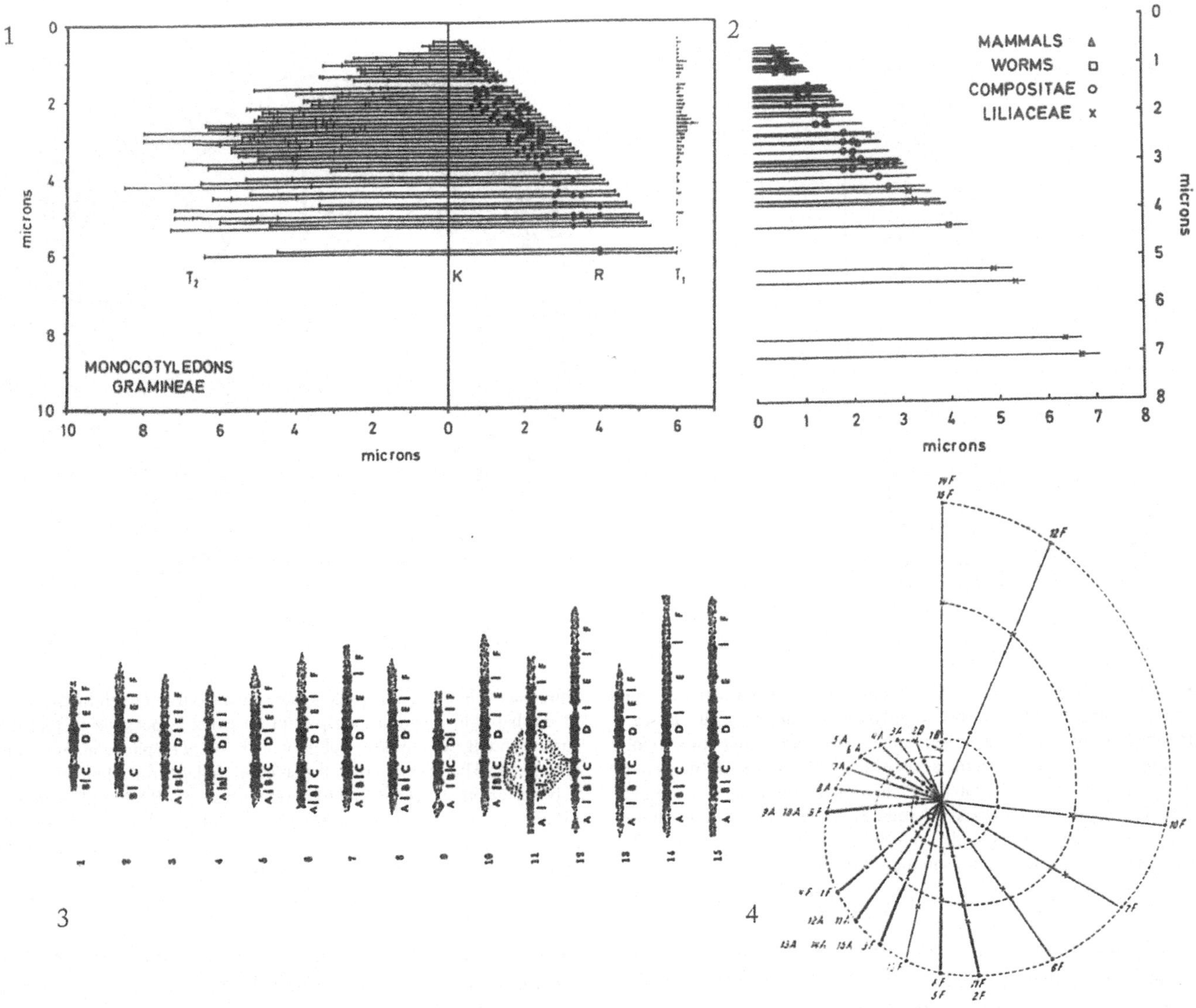

Fig. 16.5 Genes for ribosomal RNA occur in chromosomes in a regular way that allows to predict their location. Also the distribution of heterochromatin in chromosomes is distributed in non-random fashion. (**1**) Location of the ribosomal RNA genes *R* (*solid circles*) in the chromosomes of the plant family Gramineae (*grasses*). Central axis, position of kinetochore (*k*) or centromere. T_1 and T_2, telomeres at end of each chromosome arm, at metaphase of mitosis. The graph shows 149 chromosomes from 97 species. (**2**) Location of the ribosomal RNA genes in mammals (*triangles*), worms (*squares*), Compositae (*circles*) and Liliaceae (*crosses*). Only the chromosome arm containing the ribosomal genes is represented. (**3**) The distribution of heterochromatin on both sides of the centromere in the 15 chromosomes of the plant *Agapanthus umbellatus*. The large chromomeres get smaller, and the arms get less dark as they approach the chromosome ends. They build a heterochromatic gradient which is divided into sections *A* to *F*. (**4**) The distribution, of the 30 arms of the 15 chromosomes of *Agapanthus umbellatus*, according to their length building a spiral. The heterochromatic regions that they display, which increase successively in length as the arm length increases, also build spirals

not be altered although they were constantly transmitted from cell to cell and from species to species during millions of years.

Hence, two processes have worked side by side. On one hand, combination has led to improvisation. The simplest structures—quarks and leptons—were used all the way to build more complex atoms, macromolecules and organisms. On the other hand simplicity and order prevailed, becoming obligatory components and participating at every step during the increase in complexity. The ensuing products could hardly be anything else but geometric structures and it cannot be excluded that the minds of invertebrates and vertebrates had, as a consequence, to be obliged to think in geometric terms.

Sources of Figures

Fig. 16.1 (**1**) Hey, T. and Walters, P. 2003. The New Quantum Universe. Cambridge University Press, Cambridge, UK (Fig. 4.10, page 59).

Fig. 16.2 (**1**) Padamsee. H.S. 2003. Unifying the Universe. Institute of Physics Publishing, Bristol and Philadelphia, U.S.A. (Fig. 2.17, page 134). Permission of Taylor and Francis Group, *LLC*, a division of informa plc. (**2**) (1) Masterton, W.L. and Hurley, C.H. 1993. Chemistry. Saunders College Publishing, Philadelphia, U.S.A. (Fig. 6.7, page 145). (2) Hargittai, I. and Hargittai, M. 1995. Symmetry Through the Eyes of a Chemist. Plenum Press, New York, U.S.A. From: Erdmann, H. 1902. Lehrbuch der Anorganischen Chemie. F. Wieweg und Sohn, Braunschweig, Germany (Fig. 1.17, page 17).

Fig. 16.3 Cohen, J.B. 1966. Diffraction Methods in Materials Science. Macmillan, London, UK.

Fig. 16.4 (**1**) Stryer, L. 1981. Biochemistry. Freeman and Co., New York, U.S.A. (Table 26.4, page 629). (**2**) Bresch, C. and Hausmann, R. 1970. Klassische und Molekulare Genetik. Springer-Verlag, Berlin, Germany (Fig. page 243). Copyright Springer-Verlag.

Fig. 16.5 (**1**) Lima-de-Faria, A. 1976. The chromosome field. I. Hereditas 83: 1–22 (Fig. 3, page 5). (**2**) Lima-de-Faria, A. 1979. Prediction of Gene Location and Classification of Genes According to the Chromosome Field. Specific Eukaryotic Genes, Alfred Benzon Symposium 13, Munksgaard, Copenhagen, Denmark (Fig. 4, page 27). (**3**) and (**4**) Lima-de-Faria, A. 1954. Chromosome gradient and chromosome field in *Agapanthus*. Chromosoma 6: 330–370 (Figs. 3 and 4b, pages 335 and 339).

17 Self-Assembly: The Primary Source of Coherence

17.1 The Mechanism Responsible for Self-Assembly Is Independent of External Information

The most significant aspect of self-assembly from the point of view of molecular and chromosomal evolution is that the ordered aggregation of molecules, macromolecules and supramolecular systems takes place independently of external information. Enzymes, ribosomes and viruses can self-assemble due to the properties inherent to their structures. Cell membranes also possess this capacity. Phospholipid molecules contain within themselves the necessary information to assemble into bilayer systems in the absence of other sources. The amino acid sequences which build polypeptide chains also carry inherent information (Stryer 1981; Golan et al. 2005).

This is exemplified by hemoglobin, which has been studied intensively. Normal adult hemoglobin is formed of two alpha and two beta polypeptide chains. When alpha and beta chains are mixed in solution one would expect, on a random basis, that the two types of subunits would form aggregates of random numbers of subunits. This is not what happens. Under normal conditions hemoglobin molecules do not bind to each other to form four alpha, four beta, one alpha and three beta or three alpha and one beta geometric systems. Hemoglobin always appears in the form of two alpha and two beta subunits.

Another demonstration of the impressive specificity of the inherent information is furnished by an experiment in which native hemoglobins from different mammalian species were mixed under favorable chemical conditions. They did not exchange subunits in any appreciable amount to form hybrid configurations (Lehninger 1975). Therefore, randomness and external information are processes which seem to be foreign to the mechanisms which shape the supramolecular systems.

17.2 No One Believed in the Self-Assembling Capacity of Viruses and of Ribosomes

When Fraenkel-Conrat contemplated the idea of self-assembling virus particles from their constituents—the nucleic acids and the proteins—no one would believe him. How could such a complex structure as an RNA virus self-assemble? The answer came quickly; the tobacco mosaic virus, with a mass of about 40,000 kdal. and dimensions of 160 × 3,000 Å, self-assembled without difficulties. These are dimensions similar to those of the smallest chromosomes found regularly in bird genomes. When the isolated RNA is added to the isolated protein, under appropriate pH conditions, they start to assemble and the virus particles formed are fully infectious and indistinguishable from the original virus (Fraenkel-Conrat 1962). The self-assembly occurs spontaneously in solution; no additional molecular component is necessary. The process has been further studied, its specificity and speed being described by Butler (1999) and Kegel and van der Schoot (2006).

Ribosomes are the sites of protein synthesis. Of all cell organelles, ribosomes are those in which self-assembly has been studied most thoroughly. The 70S ribosome particles of the bacterium *E. coli* consist of two subunits, 30S and 50S. The 30S subunit contains one molecule of 16S RNA and 21 different proteins. The 50S subunit has a 5S RNA and 23S RNA associated with 34 different proteins. The three RNAs have been isolated and most proteins purified. Functionally active ribosomes of bacteria could be self-assembled from a mixture of the purified molecular components. All the information for correct self-assembly is present in the structure of the component molecules. It was found that the order of assembly of the proteins follows a specific sequence and that the addition of one protein enhances the incorporation of the next (Nomura 1973; Röhl and Nierhaus 1982; McGrath and Butler 1997; Liljas 2004).

A. Lima-de-Faria, *Molecular Origins of Brain and Body Geometry*,
DOI 10.1007/978-3-319-06056-9_17,

17.3 "Amorphous Materials May Be Shapeless, but They Are Not Without Order"

An extensive description of symmetry in chemical compounds was undertaken by Hargittai and Hargittai (1995). Their book called "Symmetry through the Eyes of a Chemist" covers nine chapters starting with the symmetries found in single atom configurations to the packing of atoms in crystals. Order emerges at every level of atomic organization.

After describing the properties of quasicrystals, which represent an intermediate situation to highly disordered glass material, they quote Mackay (1987) who wrote in the Journal of Non-Crystallographic Solids "Amorphous materials may be shapeless, but they are not without order". "If you look only with X-ray diffraction eyes, then all you see is translational order".

What we call amorph depends on the technology that is employed.

17.4 Amorphous Materials Have an Underlying Atomic Order: It Is Only the Degree of Order that Separates Them from Crystals

Materials in the solid state can be categorized by their structures into *crystalline*, having periodic translation symmetry, *amorphous*, without periodic and orientational symmetry, and *quasi-crystalline*, having orientational but not periodic translation symmetry phases. The distinction between these three main structures turns out to be only one of degree (L. Wang et al. 2012).

This structural classification depends on the short-range order observed in liquids and amorphous materials and its transition to long-range order evident in the periodicity inherent to crystals. Nine different types of structures can be distinguished in the transition from amorphous materials to highly crystalline structures (Wang and Fernandez-Martinez 2012). (1) *Glasses and amorphous materials*, without long-range order but hints of medium-range order. (2) *Amorphous materials* exhibiting medium-range order (*MRO*). (3) *Nanocrystals*, which have no long-range order but grow so slowly that they aggregate in a structurally coherent manner. (4) *Clay minerals and solid electrolytes* (silver cations), with a certain degree of disorder compensated by a partial crystallization. (5) *Mesocrystals* having an intermediate nonclassical type of crystallization. (6) *Crystals with structural defects*, in which atoms foreign to the structure, usually called impurities, intercalate causing structural distortions. (7) *Highly crystalline structures* which assume the highest form of coherence. (8) *Quasicrystals*, these have poor long-range ordering but a high degree of short-range rotational symmetry. (9) *Ordered amorphous carbon clusters* (*OACC*) are a crystalline material composed of amorphous clusters. X-ray diffraction and other techniques led to the discovery of a long-range ordered material constructed from units of amorphous carbon clusters that was synthesized by compressing fullerenes (large associations of carbon atoms building polyhedral molecular geometries) (L. Wang et al. 2012) (Fig. 17.1).

Hence, there is a continuum between the two ends of the structural spectrum.

17.5 Quasicrystals Are Non-Random Structures: The Impossible Became Possible

The term quasicrystal was coined by Paul Steinhardt who studied the structure of metallic glasses in 1984. The difference between a crystal and a quasicrystal is that the repetition of a unit cell is regular in the crystal whereas it is not regular, nor it is random, in the quasiperiodic structure.

These structures were discovered by Shechtman in 1982 when he did electron diffraction experiments on materials produced by very rapid cooling of molten metals. He observed an electron diffraction pattern with tenfold symmetry. This was a great surprise to crystallographers who before had considered such patterns to be impossible. As a result Shechtman waited over 2 years to publish his result (Shechtman et al. 1984) and waited still longer to become a Nobel Laureate. At present, quasicrystals have been produced at will, displaying a fivefold symmetry. This is a most common symmetry in flowers and some animals (starfish) but absent in minerals. Such a discrepancy was one of the arguments employed by evolutionists to show that the patterns of living organisms had no relation to those found in the inorganic world. Quasicrystals disposed of this vitalistic attitude.

A quasicrystal is now defined as a "highly ordered atomic structure, yet the clusters repeat in an extraordinarily complex nonperiodic pattern" (Steinhardt 1990). Recent advances in the fabrication of quasicrystals have led to obtain them from nanoparaticles and colloids (Xiao et al. 2012).

17.6 Evolution of Crystals: Nanoparticles of Different Shapes Give Rise to Different Crystal Types

The evolution of crystals has become now a general concept after it was introduced by Lima-de-Faria (1988) when no one would dare to propose evolution at the mineral level. This evolution is now being studied experimentally by various

1

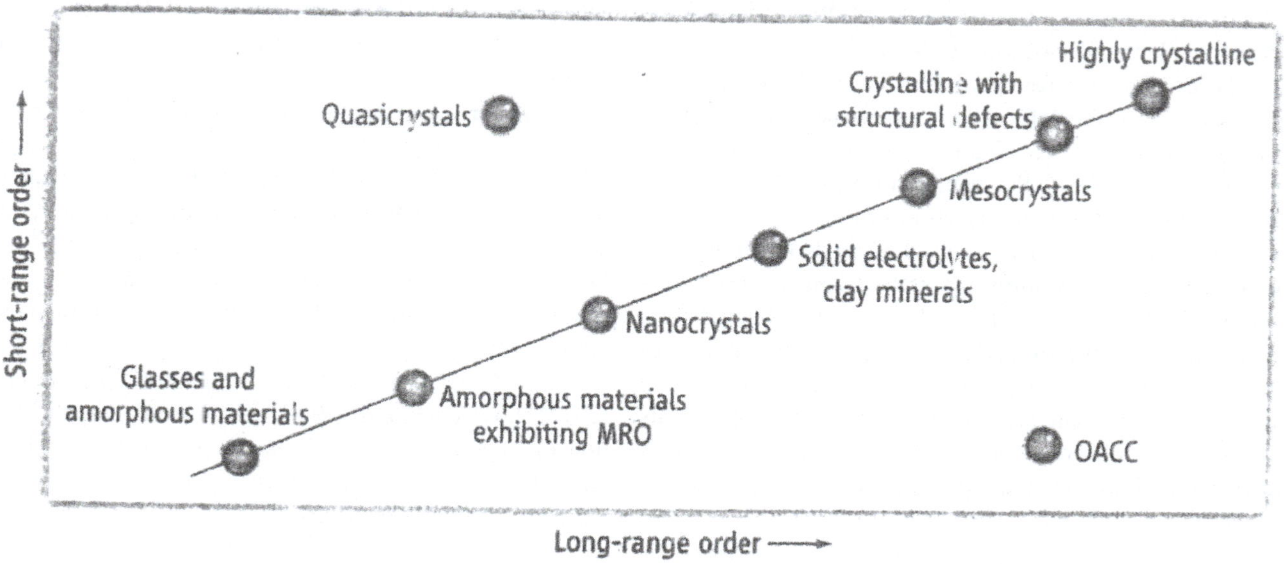

2

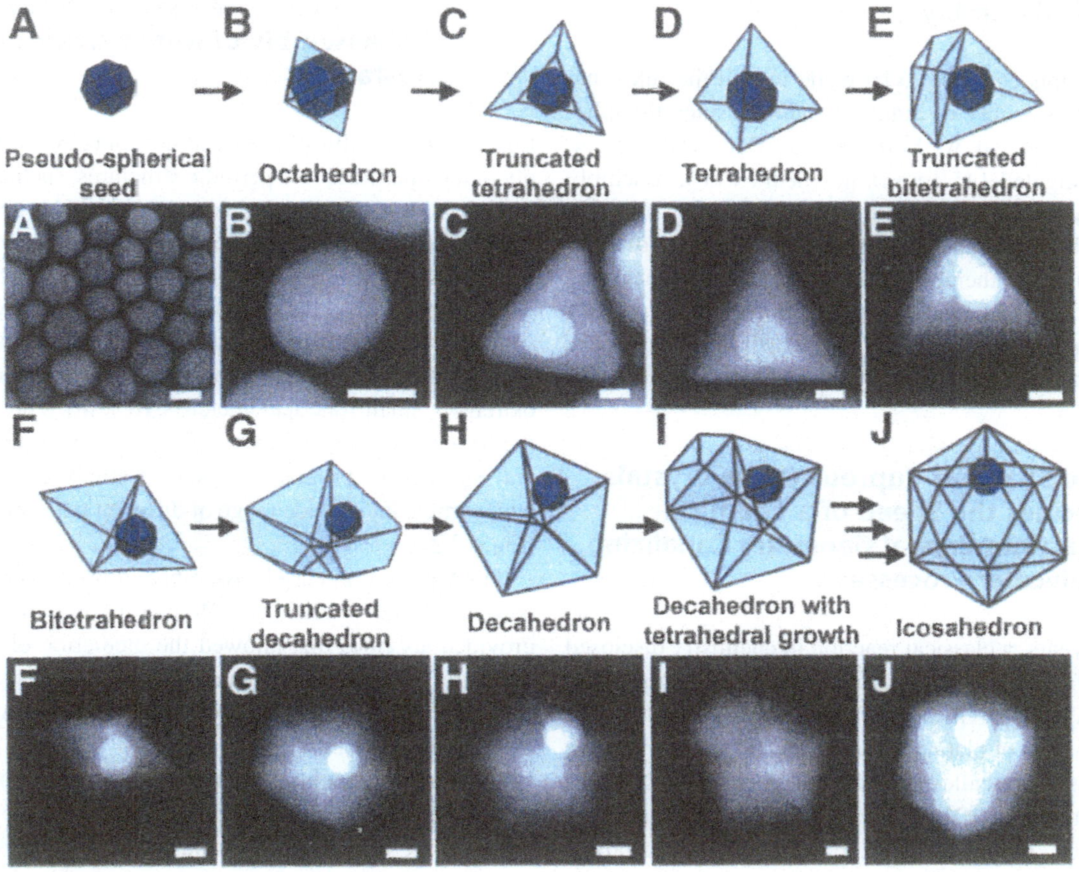

Fig. 17.1 **(1) Materials in the solid state build a continuum from *amorphous* substances to *highly crystalline* structures classified into 9 classes.** Intermediates between these forms are: *amorphous materials exhibiting medium-range order* (***MRO***), *nanocrystals*, *solid electrolytes*, *mesocrystals*, *crystals with impurities* that result in defects. *Quasicrystals* and *Ordered amorphous carbon clusters* (***OACC***) represent classes in which order is also evident but takes other forms of atomic organization. **(2) Stepwise evolution of spherical seeds into 20-fold twinned icosahedra**. Experimental demonstration of the growth of gold nanoparticles used as a seed (and as an electron microscopy label) accompanied by a second material—silver. This allowed to follow the growth of the particles into icosahedra by using scanning transmission electron microscopy. **A–J.** *Upper row*, models of electron micrographs (images shown in *lower row*). The evolution was monitored over time using fluorescence and isotopic labeling. All observed particle shapes can grow from a single crystalline nanoparticle seed

research groups (Langille et al. 2012; Y. Wang et al. 2012; Wang and Fernandez-Martinez 2012).

The growth of a nanoparticle from a single-crystalline gold seed into a bimetallic silver-gold icosahedron structure, was followed using scanning transmission electron microscopy (STEM). It led to the identification of eight intermediate structures during their crystal growth.

Gold nanoparticles can be used as seeds to grow silver shells and ultimately form bimetallic triangular prisms. A variety of different nanoparticle shapes such as bitetrahedra, decahedra and icosahedra were produced in the same reaction. All the observed particle shapes can grow from a single-crystalline nanoparticle seed. The growth of crystals can now be followed and monitored under the electron microscope (Langille et al. 2012, Fig. 17.1).

17.7 Prediction of Complex Structures Becomes Possible As a Result of Self-Assembly

Several attempts continue to be made to find the rules that regulate self-assembly of atomic structures since this phenomenon is now so well-established. Damasceno et al. (2012) investigated 145 convex polyhedra whose assembly arises solely from their directional shape. They found a remarkable high propensity for self-assembly and structural diversity which led them to show that from simple measures of particle shape, and local order in the fluid, the assembly of a given shape into a liquid crystal, plastic crystal or pure crystal can be predicted (Fig. 17.2).

17.8 The Shape of Cuprous Oxide Crystals Decides the Speed of a Chemical Reaction: Crystal Geometry Canalizing Molecular Processes

Known for years, a classical reaction in chemistry disclosed that the surface of atoms in different faces of a crystal had distinct catalytic activities. Surfaces of the type called 111 on single crystals of platinum were more catalytically active than surfaces (of so-called type 100) in reactions that formed aromatic compounds (Yang 2012). There has been a growing body of literature on surface-dependent catalytic activities (increases in the rate of a reaction) notably for metals. Huang et al. (2012) were able to study this type of reaction in a novel and most exact way. They developed a simple method to control the shape of cuprous oxide crystals, with dimensions of approximately 200–300 nm, producing nanocrystals as cubes, rhombic dodecahedra (polyhedra with 12 rhombic faces) and a variety of other geometric solids. The novel particle morphologies were carefully analysed and measured by Scanning Electron Microscopy and the reactions were followed by color changes occurring in a light-activated reaction using methyl orange.

The results were startling. The catalytic properties of the crystals were dependent on the kinds of faces that the crystals had. Cubic crystals were mainly inactive, octahedra were moderately active and rhombic dodecahedra were most active. This last crystal shape, exposing only the 110 type faces, exhibits an exceptionally good photocatalytic activity toward the fast and complete light-degradation of methyl orange (Huang et al. 2012).

These experiments as well as those of Langille et al. (2012) show how the shapes of crystals can guide the direction of organic reactions, a most significant result in the understanding of the emergence of geometric patterns in the body of living organisms.

17.9 "Self-Assembly Enters the Design Era": Simple Rules Are Used to Design the Self-Assembly of Nanoparticles into Crystals

Enormous difficulties have been encountered in the assembly of nanoparticles into periodic structures, such as crystals. The effort is worthwhile because these particles hold the key to the discovery of materials with electronic and optical properties of the utmost importance in today's technology.

Macfarlane et al. (2011) have succeeded in overcoming some of these difficulties by attaching complementary single-stranded DNA to nanoparticles as linkers. Nanoparticles are extremely small (one nanometer is 10^{-9} m). The result was the formation of crystal lattices. The linkers provided a general strategy to program the self-assembly of almost any nanoparticle into a wide range of different periodic structures (Fig. 17.2).

DNA-programmed self-assembly allowed to obtain nanoparticle crystals with nine different space symmetries and provided six rules that allowed the prediction of the lattice structure into which the nanoparticles will self-assemble. The lattice is controllable by the size of the DNA molecule and length of the DNA's sticky ends.

Nanoparticle geometry is now a general concept but there are still difficulties looming ahead. The crystalline structures reported are limited in size and how these rules apply to particles of larger sizes is an open question (Travesset 2011).

17.10 In Colloids Self-Assembly Is Largely Controlled by Particle Geometry

The ability to design and assemble three-dimensional structures from colloidal particles has been limited by the absence of specific directional bonds in these molecular

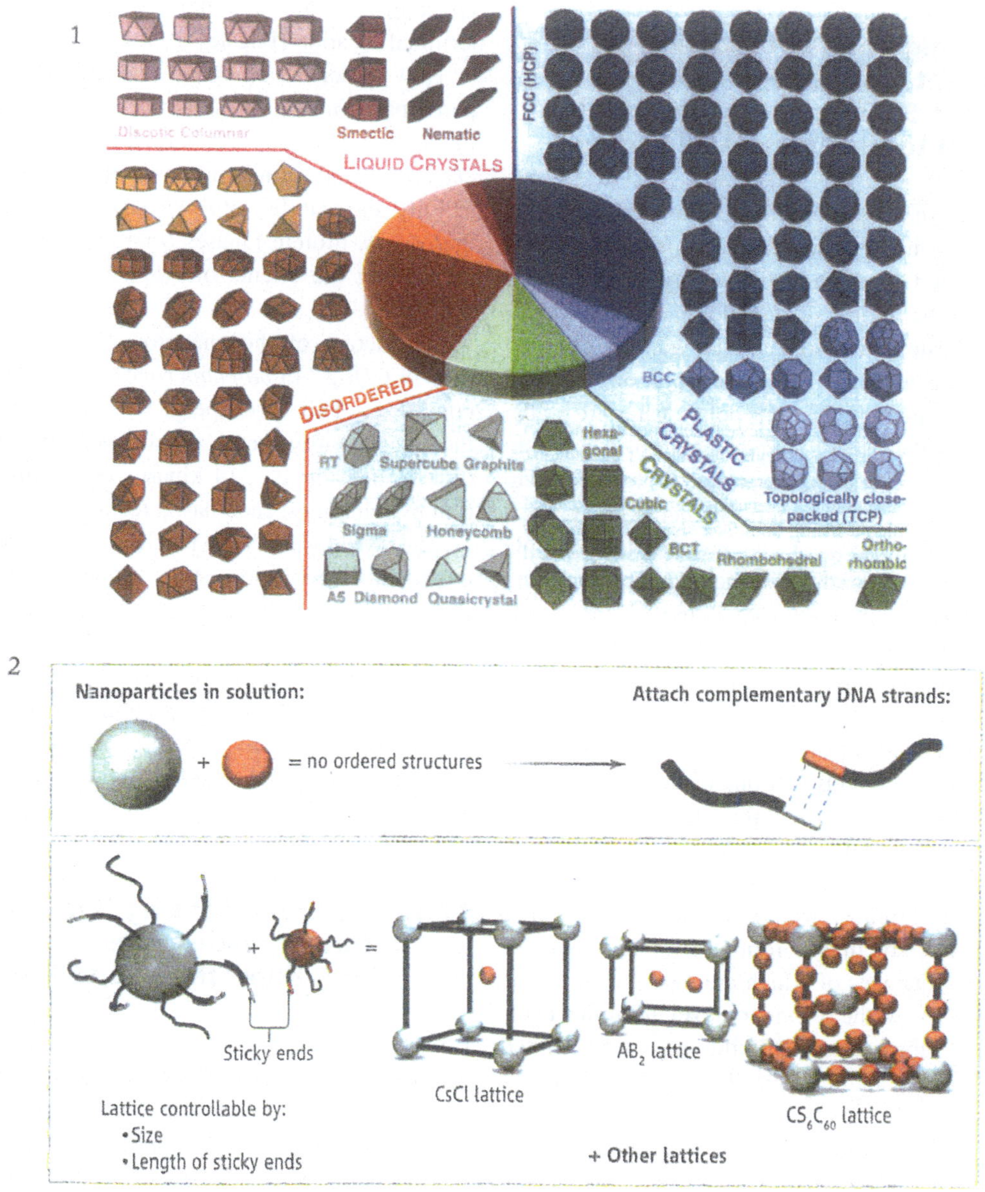

Fig. 17.2 (1) **The self-assembly of polyhedra into complex structures leads to atomic patterns that can be predicted.** Polyhedra such as crystals, plastic crystals and liquid crystals are formed from disordered fluid phases demonstrating a strong propensity for order even in the absence of attractive interactions. Polyhedra are separated into four categories of organization as indicated by different colors and disordered (glassy) phases. The *circular* chart in the center of the Figure compares the frequency of the 10 observed classes. (2) **Simple rules are available to design and control the self-assembly of nanoparticles into complex structures.** Attaching nanoparticles with complementary single-stranded DNA linkers results in controlled self-assembly of these particles. AB_2 is a lattice isostructural with aluminum diboride

edifices. Only a few simple crystals had been obtained. Y. Wang et al. (2012) developed a general method in which they fabricated patchy colloidal particles having DNA attached to them. These DNA molecules had single-stranded "sticky ends" allowing the formation of highly directional bonds.

The particles that were coated with DNA built clusters that from the beginning had different types of well-defined symmetries. Remarkable, was that the formed colloidal structures inherited the symmetries of their parent clusters as confirmed by the use of fluorescent dyes and electron microscopy. These features allowed the particles to self-assemble with triangular, tetrahedral and other geometric figures that could be programmed.

17.11 The Configuration of Proteins, so Important for Their Function, Is Assumed Spontaneously Because It Is Inherent to the Amino Acid Sequence

The X-ray analysis of the protein myoglobin (which closely resembles hemoglobin) furnished detailed information on its atomic structure and it explained how myoglobin folds. A driving force in the folding of myoglobin is the tendency of the amino acid residues to avoid cellular water. The amino acid chain folds spontaneously in aqueous solution so that its hydrophobic chains are buried and its polar chains are on the surface. Myoglobin consists of a single chain of 152 amino acids that during its folding captures a heme group containing an iron atom at its center. The heme group has a decisive role on the folding of the protein. In a simple experiment myoglobin is devoid of its heme group becoming apomyoglobin which is unfolded when in urea. If the urea is removed it partially refolds, but only when the heme group is captured spontaneously does it assume the final shape of myoglobin.

Stryer (1981) concludes that the folding is inherent to the amino acid sequence and adds that one of the challenges of chemistry is to discover the rules that determine how the amino acid sequence specifies the three-dimensional structure. Some of these rules are starting to unfold. Most proteins go through similar foldings which consist of four major levels. The primary structure is formed by long chains of amino acids, the sequences of which are determined by the coding bases in DNA. At the secondary level the amino acids interact with themselves building well-defined conformations: sheets and helices. The tertiary structure arises when amino acids more distal to one another along the chain form disulfide bridges. Later the quaternary structure occurs when independent protein subunits come together, as is the case in hemoglobin which consists of four polypeptide chains combined with four heme groups. These are located not at random but in crevices near the exterior of the giant molecule.

17.12 "Extraordinary Molecular Recognition" in Proteins that Counteracts Detrimental Mutations

The idea prevailed some decades ago, that the cell was a "soup" where molecules moved by Brownian movement and hit each other at random. This approach is long gone, but remains in the minds of some biologists.

The last 20 years have seen the demonstration that most molecules have a molecular label that addresses them to a specific site. One of the main functions of the Golgi apparatus, is the labelling of proteins specifying their terminal location (Pugsley 1989).

Now order is reaching deeper. The protein *HSP90* participates in a large number of cellular processes involved in molecular evolution and disease (cancer). It is a *molecular chaperone*, a denomination given to proteins required in the activity and stability of other proteins which are known as the *clients*. Protein kinases (that also participate in cancer) are among the clients of *HSP90* (Samant and Workman 2012).

The action of the chaperone is facilitated by another protein *CDC37* (a *co-chaperone*) creating larger and stable complexes. This co-chaperone facilitates the recognition of the kinase-family chemical features by HSP90. Taipale et al. (2012) investigated 314 kinases determining their interaction by a luminescence-based technique.

The conclusion was reached that "The extraordinary molecular recognition skills of the chaperone *HSP90*" allow this protein to interact selectively with hundreds of other proteins of diverse function. This action has the additional effect of directing evolution. The recognition mechanisms of the *HSP90* protein counteract the effects of mutations, that could be detrimental to the organism, providing improved survival under stress (Samant and Workman (2012).

17.13 Molecular Sensing Is Now a Standard Term: Proteins Sense Amino Acids and Cells Know the Size of Their Organelles

Cells have the ability to sense whether nutrients are scarce or abundant accordingly modifying their metabolism in an active way. A key molecule that senses the nutrient status in higher organisms is an enzyme *target of rapamycin* (TOR) that phosphorylates a subset of proteins functioning in cell growth and metabolism. The enzyme associates with other proteins to build the complex *mTORC1*. This has the remarkable property of sensing amino acids within the cell but the mechanism was not fully understood (Abrahamsen and Stenmark 2011) (Fig. 17.3).

Zoncu et al. (2011) showed that this multicomponent protein is a kinase, which functions as a *master growth regulator*, that is stimulated by amino acids. Sensing of amino acids inside lysosomes initiates a chain of events that stimulates cell growth.

Size-sensing mechanisms range from direct measurements provided by direct reporters to indirect functional readouts and are used to modify the cell assemblies under both normal and stress conditions. Chan and Marshall (2012) have studied these events in flagella. The length of a flagellum is dependent on competing processes of assembly and disassembly. These rely on an intraflagellar transport

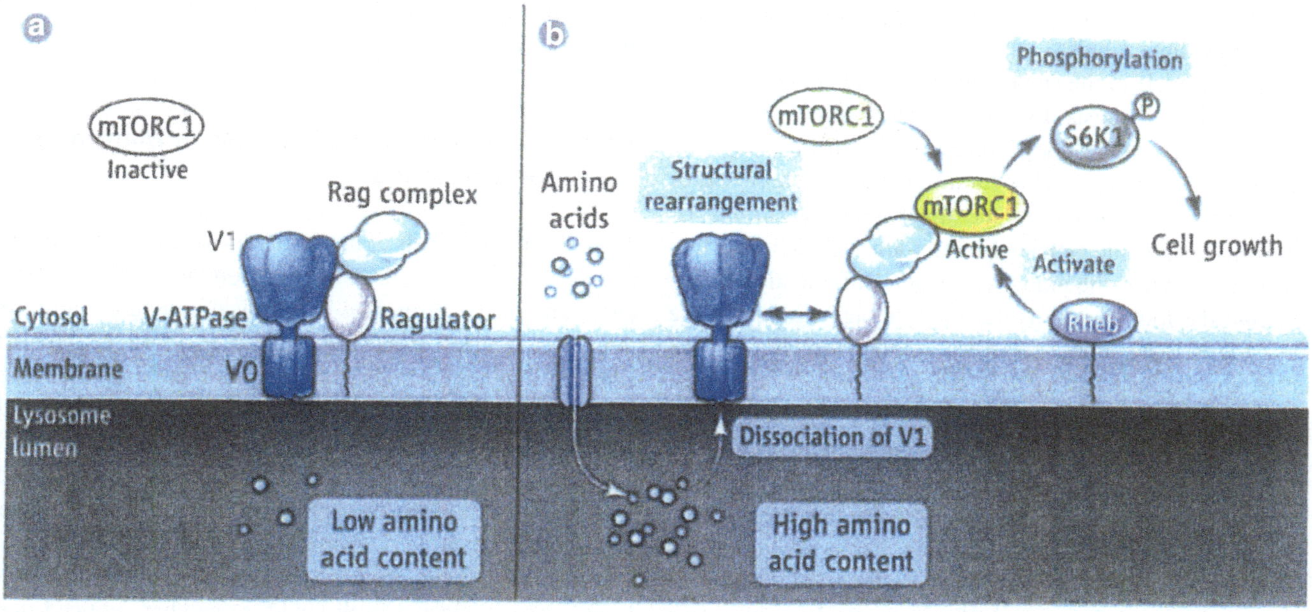

Fig. 17.3 Amino acid sensing. (**A**) At low amino acid concentrations the *Regulator* interacts with the *Rag complex* not allowing contact with *mTORC1*. (**B**) When amino acids are abundant, a structural rearrangement occurs that alters this interaction resulting in the incorporation of *mTORC1*. This chain of events stimulates cell growth

which carries *tubulin* to the growing flagellar tip. When the two flagella in the green alga *Chlamydomonas* are severed they regenerate and grow rapidly back to the original size. If only one flagellum is severed, the other intact flagellum shortens until the two flagella reach equal lengths at which point they resume growth together indicating that the cell "knows" how long both its flagella are since *Chlamydomonas* is a single cell organism.

Telomeres are DNA sequences that cap chromosomes—without them they cannot function or survive. At their ends, chromosomes contain a telomere-length control that depends on the accumulation of a telomere repeat binding factor in human cells. This is a protein that binds to DNA sequences and recruits a number of other factors. Long telomeres have more protein bound. As the chromosome reaches a certain length, the telomeres get saturated, and the chromosome becomes inaccessible to further lengthening (Chan and Marshall 2012).

Several questions remain to be answered to fully elucidate these long range molecular processes. One of them is how is the size sensor calibrated?

17.14 Organ Assembly in a Dish from Stem Cells

The pituitary gland of vertebrates regulates essential physiological processes, including growth, puberty and reproduction. This occurs under the control of the hypothalamus in the brain.

In the early embryo, the tissues that later will form the hypothalamus and *Rathke's pouch* are adjacent and remain in contact. The pouch is a precursor structure to the pituitary gland from which hormone-secreting cells can be made to differentiate. Suga et al. (2011) report the regeneration in a petri-dish of a functional embryonic organ from mouse stem cells. They actually were able to induce the formation of *Rathke's pouch* demonstrating that tissues can interact to form organs, the process being controlled and occurring in a largely autonomous manner. As Rizzoti and Lovell-Badge (2011) write, it is not surprising that the factors required in this experiment were the same as those active in the mouse and human brains including *Sonic Hedgehog*, bone morphogenetic proteins and fibroblast growth factors. During the experiment they also recognized the same cell movements that occur in the living brain.

17.15 Self-Assembly Extends from Organs to Organisms

The cells of marine sponges can be dissociated by pushing them through fine bolting cloth. The cells settle at the bottom of the vial but soon they start to aggregate by means of their amoeboid movements. After 4 days complete sponges with canals and flagellated chambers emerge. The chemical responsible for self-assembly has been isolated. This substance is species specific since it does not cause the self-assembly of sponges from other groups (Galtsoff 1923; Müller et al. 1976).

The self-assembly of a whole organism is still more evident in slime molds.

The life cycle of the amoeboid *Dictyostelium discoideum*, is very short, it lasts for only 4 days. Each individual possesses a fruiting body which, when mature, releases many spores. Under suitable conditions each of these spores opens up and hatches one unicellular amoeba. During the vegetative state the amoeboid cells are separate and free swimming. When the amoebae become sufficiently numerous and the food supply is depleted the aggregation of the cells starts. The amoebae stream radially toward central collecting points, and as they reach the center, a conical mound of cells is built up (Bonner 1947; Gerisch 1968). The cell mass elongates and contains between a few hundred to 100,000 cells. A molding and sculpturing of the cellular mass follows the aggregation process, and the wormshaped slug migrates by gliding at a speed of 2 mm per hour. This movement is due to the fact that the individual amoebae which constitute the organism are themselves in active motion. Finally, during the fruiting period, the whole cell mass rights itself and grows up into the air to form a delicate stalk supporting at its apex a round, smooth, spore body. The fruiting body consists of three distinct cell groups: spores, stalk cells and basal disc cells. These have different sizes, shapes, constitutions and functions, as do the various tissue cells of, for instance, a frog embryo (Bonner 1959).

A protein has been identified which is present on the surface of cells with the capacity for self-assembly but which does not occur on mutant cells unable to aggregate. This protein, named *discoidin*, is supposed to act as a specific cell-adhesion molecule that binds *D. discoideum* cells together by reacting with specific receptors on adjacent cells (Gregg and Yu 1975). Since then another protein, called myosin, has been found to be involved in the assembly of this organism (Kuczmarski and Spudich 1980; de la Roche et al. 2002).

17.16 Self-Assembly of Tissues in Mammals and Humans

The cells of the gonads of newborn rats were dissociated into a suspension of free cells. Under appropriate conditions both testes and ovary reorganize into typical structures from the free cells (Zenzes et al. 1978; Ohno et al. 1978). Human capillary endothelial cells from long-term cultures were also able to form tubular networks in vitro. Under the light and electron microscope, these appear almost identical to capillary vascular beds found in vivo (Folkman and Haudenschild 1980). Human skin has been produced by allowing dispersed cells to self-assemble (Dubertret et al. 1987).

When specimens of the African spiny mouse (*Acomys kempi*) were collected in the wild, researchers noticed that during capture the animals skimmed away by losing 60 % of the skin of their back. Seifert et al. (2012) report that the animals regenerated the skin with hair within 30 days of damage. They further tested the regenerative capacity of *Acomys* tissue by having a hole punched through the ear. They found that soon the hole was filled by dermal, epidermal and cartilage tissues but not by muscle, a situation comparable to what happens during complete limb regeneration in salamanders. Three molecules: *fibronectin*, *tenascin* and *collagen III* participated in this process.

The self-assembly can be traced all the way to the formation of human tissues. The properties of the atoms and molecules could not be erased because they are inherent to their internal construction.

Sources of Figures

Fig. 17.1 (**1**) Wang, D. and Fernandez-Martinez, A. 2012. Order from disorder. Science 337: 812–813 (Fig. page 812). (**2**) Langille, M.R. et al 2012. Stepwise evolution of spherical seeds into 20-fold twinned icosahedra. Science 337: 954–957 (Fig. 1, page 955).

Fig. 17.2 (**1**) Damasceno, P.F. et al. 2012. Predictive self-assembly of polyhedra into complex structures. Science 337: 453–457 (Fig. 1, page 454). (**2**) Travesset, A. 2011. Self-assembly enters the design era. Science 334: 183–184 (Fig. page 183).

Fig. 17.3 (**1**) Abrahamsen, H. and Stenmark, H. 2011. Growth signaling from inside. Science 334: 611–612 (Fig. page 611).

Part V

The Brain Has Its Own Geometry and Is Able to Change the Body's Geometry in Seconds

The Presence of Chromosomes Is Not Obligatory for Brain Function: Brain, Eye, Blood and Skin Cells Work Efficiently Without a Nucleus 18

18.1 A Fully Functional Brain, with Neurons that Lack Chromosomes, Is Found in a Wasp

Decrease in body size has occurred in many animal groups during evolution but is most evident in insects where their organs have suffered alterations at the the cell level.

The smallest flying insects belong to the genus *Megaphragma* (Hymenoptera). The adult body length of these wasps is only 170 μm in *M. caribea* and 200 μm in *M. mymaripenne*. They are of the same magnitude as single-cell organisms like *Paramecium* and *Amoeba* (Fig. 18.1).

Animals collected in Portugal, Spain and Ukraine were analysed with the light and electron microscopes and the number of neurons was estimated from serial sections (Polilov 2012).

The central nervous system (of which the brain is the largest structure) displays in *M. mymaripenne* thoracic, metathoracic and abdominal ganglia. During early development (pupal stage) the central nervous system of *M. mymaripenne* has about 7,400 cells and a structure typical of other insects. The brain contains only 4,600 cells. As pupae approach their final phase there is a mass lysis of neurons. As the adult stage is reached the brain shrinks, becoming about half the size in adults. In this process 95 % of the neurons lose their nuclei and cell bodies but the brain continues to function normally although one is dealing with an organism with an almost entirely anucleated nervous system (consisting of circa 7,000 anucleated cells). In such cells there are no chromosomes and as a consequence no genes are present that can be expressed by producing RNAs and other molecules but some RNAs may be retained. Animal behavior is well-known to depend on the function of the brain. These minute parasitic wasps (with a lifespan of 5–9 days) sustain all life functions: feeding, flight, mating and the search for hosts where they lay their eggs as the good parasites that they have learned to be. Flight is also a function as complex as parasitism because it demands a perception of air currents, sense of direction, equilibrium and decisions controlling speed. The brain of these wasps functions perfectly, otherwise they would not be able to reproduce and sustain a normal life.

18.2 Brain Size and Its Relation to Body Volume Does Not Elucidate Animal Behavior in Insects

The brain of the wasp *Megaphragma myramipenne* contains around 4,600 neurons—orders of magnitude fewer than house flies (340,000 neurons), bees (850,000) or even related parasitic wasps (37,000). An insect performs all its vital tasks with a brain that is more than 100 times smaller than its relatives belonging to the same animal group. The relation of the brain to body volume is also relevant. In *M. myramipenne* it is 2.9 %, in honeybees (*Apis*) 0.35–1.02 %, and in ants 0.57 % (*Formica*) (Polilov 2012). As it will be seen later, in mammals this relationship does not seem to explain differences in animal behavior.

18.3 How a Cell Functions for 100 Years Without Chromosomes: Eye Lens Cells Lack Most Cell Organelles

We tend to think that cells to function must have: (1) A nucleus with its content of chromosomes consisting of arrays of genes and other DNA sequences. (2) An endoplasmic reticulum, with its many membranes, where the major cell chemical events take place. (3) Mitochondria with their respiratory functions. (4) Golgi apparatus, where proteins get a chemical tag that allows them to find the correct address in the cell. (5) Other minor structures that are also relevant to its digestive function such as lysosomes.

There are, however, cells which function normally for the whole life of a human being, without possessing all these vital cell organelles. They are the many cells that build the lens of the human eye.

A. Lima-de-Faria, *Molecular Origins of Brain and Body Geometry*,
DOI 10.1007/978-3-319-06056-9_18, © Springer International Publishing Switzerland 2014

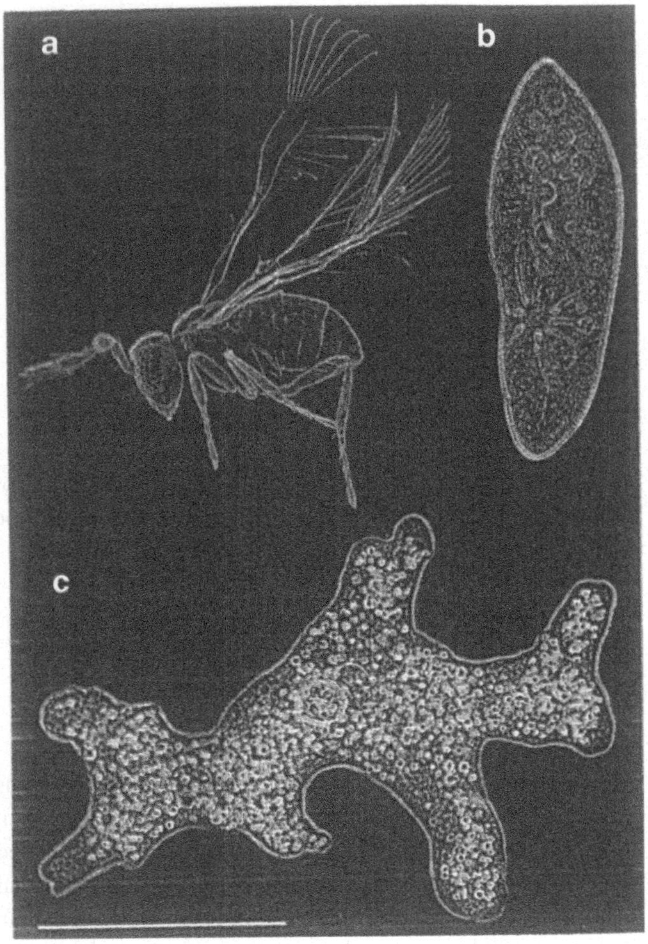

Fig. 18.1 **The wasp *Megaphragma mymaripenne*** (**A**) compared with two species of protozoa: *Paramecium caudatum* (**B**) and *Amoeba proteus* (**C**). All on the same scale. The bar is 200 μm

During vision, the light reflected by an object passes through the cornea (a transparent sheath across the front of the eye) and a fluid (aqueous humor) on through the pupil of the iris and into the lens which is normally clear. From there the light travels through the gel-like vitreous body to the retina. This last structure is the part of the eye that converts light into electrical signals that are transmitted to the brain for interpretation.

How could light pass through this sequence of structures if they were opaque? Instead they are all transparent but how have they become clear remaining at the same time fully alive and active?

The lens of vertebrate eyes is a transparent structure constructed of numerous layers of fibers rich in the protein *crystallin*. They are arranged like the layers of an onion. The fibers derive from epithelial cells that with time elongate into ribbons, but the most important cellular event is that they lose their nuclei (Koretz and Handelman 1988). The full transparency of the lens is subsequently achieved by the destruction, not only of the chromosome set, but of the other major organelles that it originally contained.

The mechanism of organelle degradation is a programmed process. Newly formed fiber cells are layered into existing fibers. As a result the lens contains cells at all stages of terminal differentiation. In the innermost (oldest) fiber cells the nuclei are already absent, but in the superficial fiber cells, the nuclei are still ovoid, assuming a more spherical shape prior to their disintegration. DNA synthesis is arrested and high concentrations of crystallin proteins accumulate (Counis et al. 1998).

Disintegration of the chromosomes in bovine and chicken lenses begins after this stage. It turns out that lens fiber cells are transcriptionally active until quite late in differentiation transcribing novel messenger RNAs and producing proteins. Transcription is shut down in these cells shortly before their differentiation. Only crystallin messenger RNAs are detectable several weeks after nuclear disintegration.

Unlike dying cells of other tissues that go through so called apoptosis, the lens fiber cells differentiate following a "highly ordered pattern of temporal progression" (Counis et al. 1998) and persist throughout the life time of the organism (Wride 2000). Another difference is that the cytoskeleton remains intact during organelle loss in lens fibers, while in apoptosis proper it is degraded.

Surprising is that the nucleoli, which were attached to the chromosomes at the site of the ribosomal RNA genes, were allowed to remain intact. The loss of mitochondria, from bovine, chicken, rat and monkey lens fibers, is abrupt, occurring within 2–4 h, and coincides with the disappearance of the nuclei and the endoplasmic reticulum. The Golgi apparatus disintegrates earlier being already absent when the mitochondria and the nucleus start to disappear. This means that the succession, by which not only the nucleus but also the other different cell organelles are eliminated, is an ordered event (Wride 2000).

18.4 Human Red Blood Cells Lack Nucleus, Mitochondria and Internal Membranes

Most blood cell types have a nucleus, the well-known example being the white blood cells, or leucocytes, which take many forms and perform vital functions but which lack haemoglobin.

Haemoglobin is a protein respiratory pigment with iron, which binds oxygen reversibly, involved in the uptake and transport of oxygen during respiration. When this macromolecule combines with oxygen it becomes scarlet—the red blood cells deriving their name from this colour.

The red blood cells are generated in the bone marrow from cells which are already non-nucleated but which develop from nucleated ones. Step by step the red blood cells finally lose their nucleus, mitochondria and most internal membranes (Abercrombie et al. 1990).

Ji et al. (2011) have described this transformation in detail.

This uniquely mammalian process is a rigidly programmed genetic event. A cell type called proerythroblast has initially a large nucleus that starts to condense transforming the cell from early to late erythroblast. The chromatin condensation is orchestrated by a specific microRNA (*miR-191*) that targets a series of genes (*Mxi1*, *Riok3*) which in turn regulate the activity of other genes *c-myc* and *Gcn5*. Chromatin condensation also involves the histone protein which is intimately combined with the chromosome's DNA. These events change the transcription activity of this last molecule. A final step leads to the extrusion of the nucleus for which a contractile actin ring is necessary. The final product is the red blood cell (RBC) which has a basic function in the human body but which lacks chromosomes.

Again, an ordered molecular cascade led to the removal of cell organelles by a process directed from within the cell.

18.5 Programmed Transformation of Skin Cells Results Also in DNA Degradation

As the skin forms, during the terminal differentiation of the epidermal keratinocytes the nucleus undergoes a programmed transformation from an active state, associated with cornification and epidermal barrier formation, to a fully inactive condition characteristic of the keratinized cells of the cornified layer of the skin. As a final stage, DNA degradation occurs in these cells.

The molecular events involved have been studied in detail by Botchkarev et al. (2012). A series of genes participate in the regulation of gene expression in the epidermal differentiation complex. Histone acetylation is also part of the program which leads to repressed transcription and, as a consequence, inactivation of genes. The final result is chromosome disintegration in cells that give rise to nails, hair follicles, glands and other skin formations of which the protein keratin is a major component.

Brain, eye, blood and skin cells change their organization, and their function, following practically the same process.

Sources of Figures

Fig. 18.1 Polilov, A.A. 2012. The smallest insects evolve anucleate neurons. Arthropod Structure and Development 41: 29–34 (Fig. 1, page 30).

19 The Ordered Origin of the Brain

19.1 Protozoa Have No Brain but Have Features Common to Humans

Organisms which consist of one cell, the protozoa, do not possess the organs that emerged in multicellular species. They have no heart, no stomach and no brain. Yet they already contain, within the single cell, rudimentary structures which later in evolution resulted in large and highly specialized organs. These organelles, as they are called, may be sophisticated.

The digestive system of protozoa is differentiated into an oral region, a pharyngeal basket, oesophageal fibres, food vacuoles and an anal pore.

No brain is present, but a neuromotor system is at work which consists of nerve-like fibrils.

However there is already a recognizable eye. Several species, such as *Euglena* have an eye which is differentiated into a short rod, a cortical layer and an amyloid lens that allows the organism to orient itself in relation to light (Kudo 1971).

Many other features of single cell organisms are like those of vertebrates. Their chromosomes are indistinguishable from those of humans in shape and organization when seen in the light and electron microscopes. Their DNA is associated with histone proteins and a nucleolus is already present (Leick et al. 1979). Genes have been conserved between protozoa and humans to a large extent and the total number of genes is not very different. *Chlamydomonas*, another unicellular organism, has a gene number of 15,143 whereas that of humans is considered at present to be about 20,000 (Merchant et al. 2007).

Tetrahymena is an extensively studied protozoa in which Cech and Bass (1986) described the occurrence of the split gene which is a general feature of gene organization in vertebrates. Thus, although the same genetic processes are at work, in protozoa only nerve fibrils are present. No brain is in sight but the eye, in a rudimentary form, is already a functional and distinct organelle.

19.2 Vision with No Nervous System in Sponges

Sponge larvae (multicellular organisms) are able to detect light despite lacking a nervous system. These mainly marine animals are primitive in their lack of organs including mouth and gut. Besides, they do not possess the photosensitive *opsin* proteins found in all other animal eyes.

An intermediate step in the evolution of the eye is the sponge *Amphimedon queenslandica*. The transformation of this organ in other animals has always involved opsins, whereas in this sponge two cytochrome genes were identified instead. One of them is expressed in the eyes of the larva. These animals detect different light intensities and use their eyes in directing swimming.

19.3 The Search for the Origin of the Brain Has Been Based on "Severely Biased" Ideas

The question of how the human brain differs from that of chimpanzees and gorillas, as well as from other animals, has been intensely debated since evolution was established as a biological phenomenon in 1809 by Lamarck. Anatomical differences, as well as size differences, have been the main source of a controversy which ranged between 1840 and 1940. Antiquated arguments continue to be used to this day. As Striedter (2004) states the analysis of the origin of the vertebrate brain remains unresolved because its interpretation is "severely biased".

As an example, he points out the different types of phylogenetic trees that have been conceived throughout time. He considers that: "The most insidious idea in the study of brain evolution is the very old notion that biological evolution proceeded in a linear and progressive manner, from lower to higher forms of organization and with *Homo sapiens* at the very top of the so called phylogenetic scale". This

A. Lima-de-Faria, *Molecular Origins of Brain and Body Geometry*,
DOI 10.1007/978-3-319-06056-9_19, © Springer International Publishing Switzerland 2014

linear and progressive idea of brain evolution dominated throughout the 19th and most of the twentieth century since medically oriented neuroscientists and anthropologists continued to support it (Campbell and Hodos 1991). A second type of approach has concretized in the "Traditional Tree" in which humans are at the top and other animal groups are placed as side branches. A third form depicts vertebrates distributed in a "Bush or Tumbleweed Tree" in which it is stressed that nonhuman lineages do not represent blind alleys but are only the outermost terminal branchlets of this phylogeny.

"Cladograms" are a fourth solution now generally used by evolutionary biologists placing both living and extinct species along the top of a diagram. Such cladograms have also been much criticized because they lead to quite different conclusions depending on what kind of molecular information is considered most relevant (Carroll 1988).

19.4 Evolution of the Vertebrate Brain Has Been Characterized by Permanence of Organization

Enlargement of the brain has occurred multiple times during the evolution of invertebrates and vertebrates. Large and more complex brains appeared suddenly in octopus and squid as well as in insects.

In vertebrates there has been a conservatism of the major parts of the brain. The evolution of the thalamus—the ovoid ganglionic mass on either side of the brain—has been studied in four-leg vertebrates. The dorsal thalamus consists of two divisions: the lemnothalamus and the collothalamus. Their evolution can be followed starting from the ancestral stock leading to amphibians, early reptiles, early mammals, birds and modern placental mammals. The two different divisions increased in size during evolution but their different components can still be recognized along the whole evolutionary process indicating a conservatism of organization (Butler 2012) (Fig. 19.1).

19.5 Brain Size Increased Independently of Body Weight in Most Vertebrates

Manta rays have brains that are almost 40 times larger than the brains of other cartilaginous fishes with similar body weights. Also, within bony fishes (teleosts), brain size increased several times. Among mammals, relatively large brains evolved in primates, toothed whales and elephants. Humans have the largest brains among vertebrates but absolute brain size is actually much greater in elephants and in many whales than in humans (5–10 kg for whales and elephants, 1.4 kg on average for humans). When calculating relative overall body size human brains do not reach to the "top of the scale" because the brain comprises only 2 % of the body's weight in adult humans and nearly 10 % in adult mice.

It is frequently overlooked that these increases have been accompanied by major changes in brain organization. For instance, the forebrain of birds is more complex than that of reptiles. The available data have led biologists to the conclusion that brain size and brain complexity have increased several times independently.

19.6 Absolute and Relative Brain Size Not Only Increased but Also Decreased in: Birds, Bats, Elephants, Primates and Humans

Primates are placental mammals that include: lemurs, lorises, tarsiers, gibbons, great apes and humans. The variation in brain weight among living species extends from 1.8 g (*Microcebus murinus*) to 1,330 g (*Homo sapiens*). When related to body mass, primates have relatively large brains compared to most other mammals.

To elucidate this situation Montgomery et al. (2010) used 37 extant and 23 extinct primate species to measure absolute brain mass, absolute body mass and relative brain mass.

Their results confirmed that brain expansion began early in primate evolution and that increases occurred in all major groups. However, decreases in brain mass also took place along other branches, showing that in species such as *Homo floresiensis*, a human fossil recently discovered in the island of Flores (Indonesia) that lived between 95,000 and 14,000 years ago, this trend had been reversed.

Not only in humans, a reduced brain size has also appeared in living species of Orangutan and Old World monkeys. Absolute brain mass decreased in approximately 14 % of all primate branches.

This reversed process has also taken place in other vertebrates. Decrease in both absolute and relative brain size has occurred in birds, bats, bovids, elephants and hippopotami (Montgomery et al. 2010). Hence, we are dealing with a general event.

19.7 Evolution of the Different Parts of the Mammalian Brain

Two questions about brain evolution have been asked: (1) The degree to which the different functional brain systems evolve independent of each other or are dependent on the evolution of the brain as whole. (2) How phylogenetic differences in brain:body ratios have evolved.

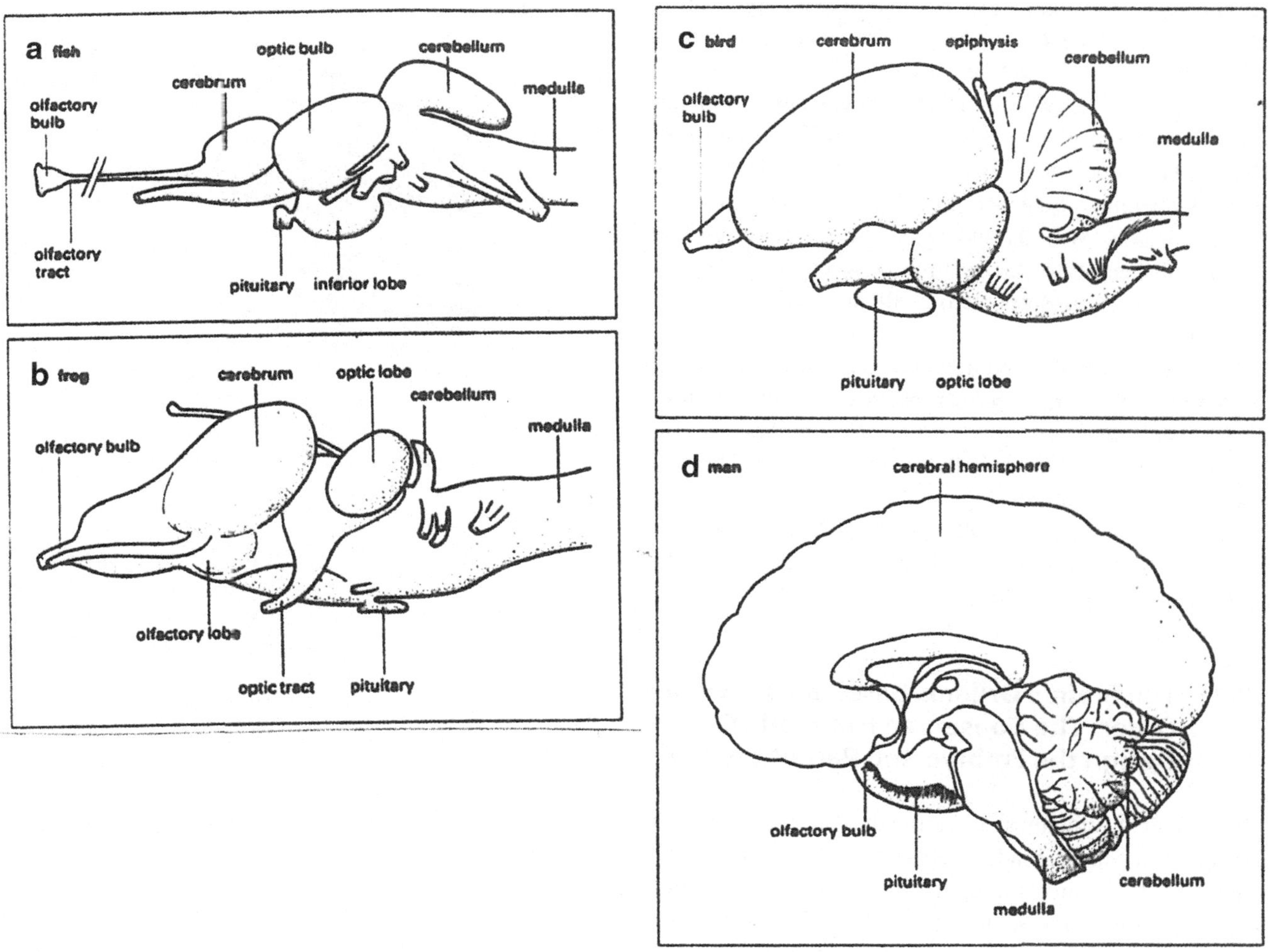

Fig. 19.1 Development of the brain during vertebrate evolution from fish to humans. The cerebrum becomes ever more important and the relative contribution of sensory areas varies according to lifestyle

The first question has led to different interpretations. The mammalian brain consists of distinct parts that fulfil different functions. Hager et al. (2012) consider a mosaic evolutionary process in which the size of the different systems evolves independently. In contrast Finlay and Darlington (1995) and Finlay et al. (2001) have proposed that the different regions are canalized by developmental factors and as a consequence evolve as a whole. Mosaic evolution assumes that heritable variation of the size of different brain parts is modulated by independent genes. On the other hand, in concerted evolution sets of genes modulate the size of multiple parts of the brain. At present the two interpretations remain to be tested at the molecular level.

The second question concerning brain evolution led to the study of 10,000 mice. It concentrated on the correlation for relative size of brain parts such as: cortex, striatum, olfactory bulb, hippocampus and cerebellum by comparing them with brain and body weights. The experiment was followed by scanning the mouse genome for genes that affected variation in body size compared with those for overall brain size. The evidence pointed to independent genes governing brain and body size.

19.8 Over 100 Species of Mammals Show that Sizes of Brain Components Are Highly Predictable As a Result of Order in Neurogenesis

Finlay and Darlington (1995) and Finlay et al. (2001) analysed data collected from 131 species of mammals including: primates, bats and insectivores. The sizes of brain components, from medulla to forebrain, turned out to be highly predictable from absolute brain size.

Contrary to expectation based on hypotheses assuming the action of selection, it turned out that there is order in the genesis of the brain across a wide range of mammals as a result of the conservation of its internal organization. This is based on a correlation between the relative enlargement of structural components as brain size increased during evolution.

The sizes of 10 measured brain subdivisions from the 131 species were plotted as a function of total brain size. These subdivisions were: Neocortex, Striatum, Diencephalon, Cerebellum, Schizocortex, Hippocampus, Septum, Mesencephalon, Medulla and Paleocortex. The result was a clear cut linearity of the relation between brain size and the 10 subdivision sizes across all species investigated (Fig. 19.2).

Schoenemann (2001) plotted brain size against body size in hominids. There has been an increase in brain size from *Australopithecus africanus* to Neanterthal *Homo neanderthalensis*. Unexpectedly, the brain size has become smaller in *Homo sapiens* (Fig. 19.2).

As Friday and Ingram (1985) point out "There is no clear relation between absolute or relative brain size and intelligence, technology or culture".

19.9 The Brain's Order Is Based on a Previous Order: The Conserved Homeotic Genes Design the Embryo and Part of the Brain

All biological structures are formed as a result of well-defined molecular cascades, otherwise a heart, or a brain, would not be formed throughout embryonic development following exactly the same path, and would never appear in the next generation having the same body location, shape and physiological activity.

Now that this chain of events starts to be unveiled, it becomes so evident that one does not think it could be otherwise.

Unexpectedly, the genes that direct the ordered segmentation of the embryo at its initial stage of development are the same that later determine the structural order present in the mammalian brain. The homeotic genes, also called *Hox* genes, have been found in the last two decades to be the determinants of the body division into segments in flies, humans, worms, plants and other organisms in which they were investigated. These genes are aggregated in a cluster along the chromosome's DNA built by about 180 base pairs. They have the following main properties. (1) They are of general occurrence, deciding the segmentation of the vertebral column in humans as well as the separation into sepals and petals in a flower. (2) Their order within the cluster has been preserved throughout millions of years, the Hox cluster of a fly being homologous to that of a human. (3) The anterior-posterior sequence along the DNA of the genes that build the cluster, corresponds directly to the anterior-posterior location of the body parts that these genes control. Hence, the order present in the embryo and the adult can be traced directly to the order occurring within the large

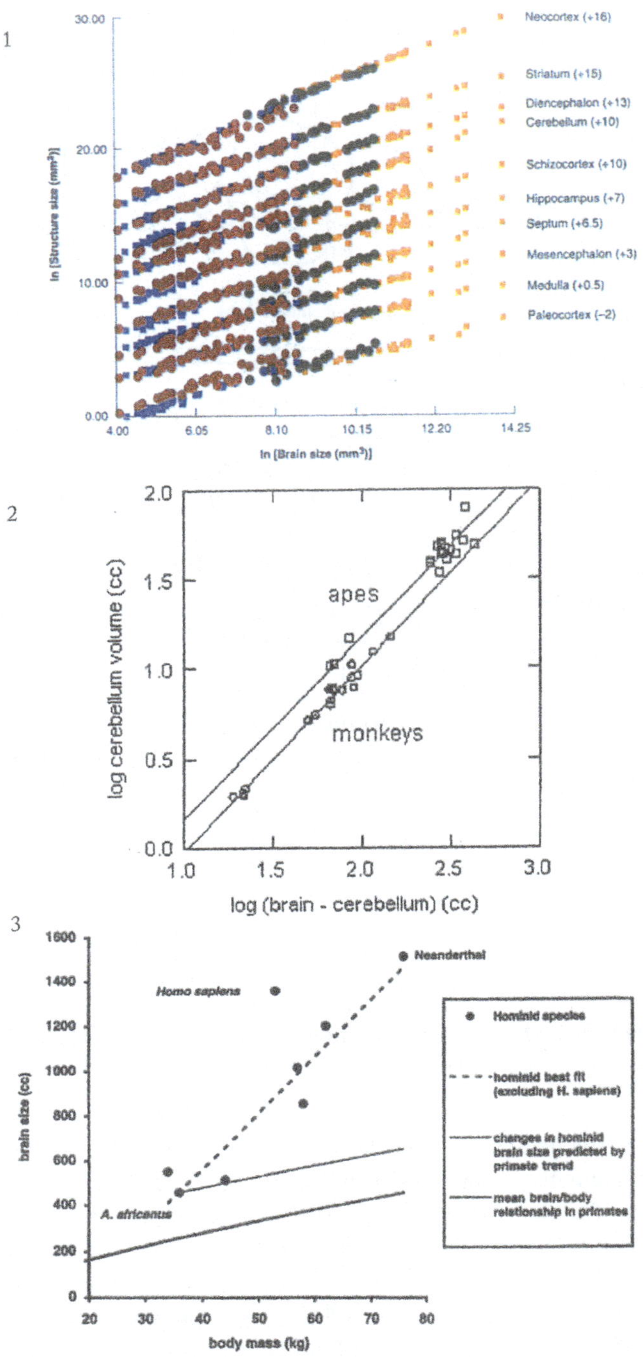

Fig. 19.2 **(1) Sizes of 10 measured brain subdivisions from 131 species plotted as a function of total brain size** (*orange squares*, simians; *green circles*, prosimians; *red circles*, insectivores; and *blue squares*, bats). This method of representation emphasizes the linearity of the relation between brain sizes and structure sizes across mammalian groups on logarithmic scales. Each scatterplot of data points corresponds to a brain subdivision. Arbitrary constants (in parentheses after each subdivision name) were added to separate the plots visually. **(2) A plot of cerebellar volume against brain volume among anthropoid primates. (3) Brain size plotted against body mass in hominid species.** *Dark solid line* represents the best fit (*least squares*) regression for primates. The *gray solid line* represents increases in brain size from *A. africanus* that would follow the primate trend. *Dotted line* represents the relationship reported by Finlay et al. for this data. The extent to which the *grey* and *dotted lines* diverge is a measure of how poorly body size predicts brain size evolution in hominids

sequence of DNA. The *Hox cluster* of Drosophila becomes, in humans, divided into four separated DNA segments *Hox A, B, C* and *D*, but in all of them the order of the genes within the cluster and their relationship to the body axis is maintained (Fig. 19.3).

Important is that the *Hox* genes also decide the formation of part of the brain.

Like in other vertebrates the A to D *Hox* genes are present in birds. The study of the brain development in the chicken revealed that a combination of *Hox* genes and related genes, are expressed in rhombomeres, a series of seven transient bulges in the developing rhombencephalon corresponding to cranial sensory and motor nerves. They are known from the chick as well as mammals.

Rhombomeres are sites of differential cell proliferation and differential cell mobility. Later in development, the pattern of axon outgrowth from the cranial motor nerves also follows the earlier rhombomeric pattern determined by *Hox* genes. This is in agreement with mutations of *Hox* genes in mice that alter the position of specific cranial nerves or prevent their formation.

In the zebrafish gene expression patterns clearly define rhombomeres. Again, mutations in these genes disrupt region-specific neural differentiation in the hindbrain (Zhang et al. 1993; Lumsden and Keynes 1989; Narita and Rijli 2009; Purves et al. 2012).

19.10 *Hox* Genes Regulate Also the Axial Skeleton

Mouse embryos have allowed to establish the different domains of *Hox* gene expression during segmentation. Several mutants of these genes created defects which helped to demarcate their areas of action.

The vertebral column, with its vertebrae and ribs, arises from the somites, which also result from the action of *Hox* genes. There are unique profiles of *Hox* gene expression within each segment which are also dependent on the *Notch* signalling cascade. Transplantation experiments and mutations in mice reinforced the evidence on *Hox* gene induction and formation of the axial skeleton. In some cases genes share the same function in patterning a certain somite.

Hox gene patterning of the rhombomeres and somites shares "fundamental features" such as colinearity and a code (Alexander et al. 2009).

19.11 Retinoic Acid, so Important in Forming the Eye, Patterns the Hindbrain

Evidence from in vitro and in vivo experiments disclosed that *retinoic acid* initiates early *Hox* gene expression and patterning of the hindbrain. In vertebrates, adding this acid during early embryogenesis resulted in the expansion of the posterior hindbrain, whereas reduction of *retinoic acid* led to an expansion of the anterior hindbrain. This acid directly activates some *Hox* genes such as *Hoxb1* and *Hoxb4* (Alexander et al. 2009).

As described in Chap. 8 retinoic acid has a primary role in both eye and body development. The fact that retinoic acid also influences the patterning of the hindbrain shows how eye, brain and body development are closely connected.

19.12 Neurons and Pigment Cells Originate from the Same Cell Population: Neural Crest Cells—It Is Thus Not Surprising that the Brain Can Alter the Pigmentation Pattern

It turns out now that the relationship between neural activity and pigmentation is not a late event but that neurons and pigment cells originated from the same embryonic tissue. This means that their interaction is much more intimate from the beginning than was originally expected.

The neural crest is a band of embryonic vertebrate ectoderm located on both sides of the developing neural tube. Neural crest cells often attain their final positions after lengthy migrations. They give rise to a series of cell types which include the spinal ganglia, glial cells of the peripheral nervous system but also melanocytes, the cells which synthesize melanin pigments (Liu and Xiao 2011). Molecular studies in mouse, chick, frog and fish have led to the division of these tissues into four major groups: (1) The cranial. (2) Part of the nervous system and heart. (3) The trunk. (4) The sacral (which forms melanocytes and sensory ganglia). Disturbances in the differentiation of these cells, due to mutations or lack of certain *Hox* genes, cause grave human diseases. For example one mutation affects normal development of pigment cells resulting in hypopigmentation, a well known human disease.

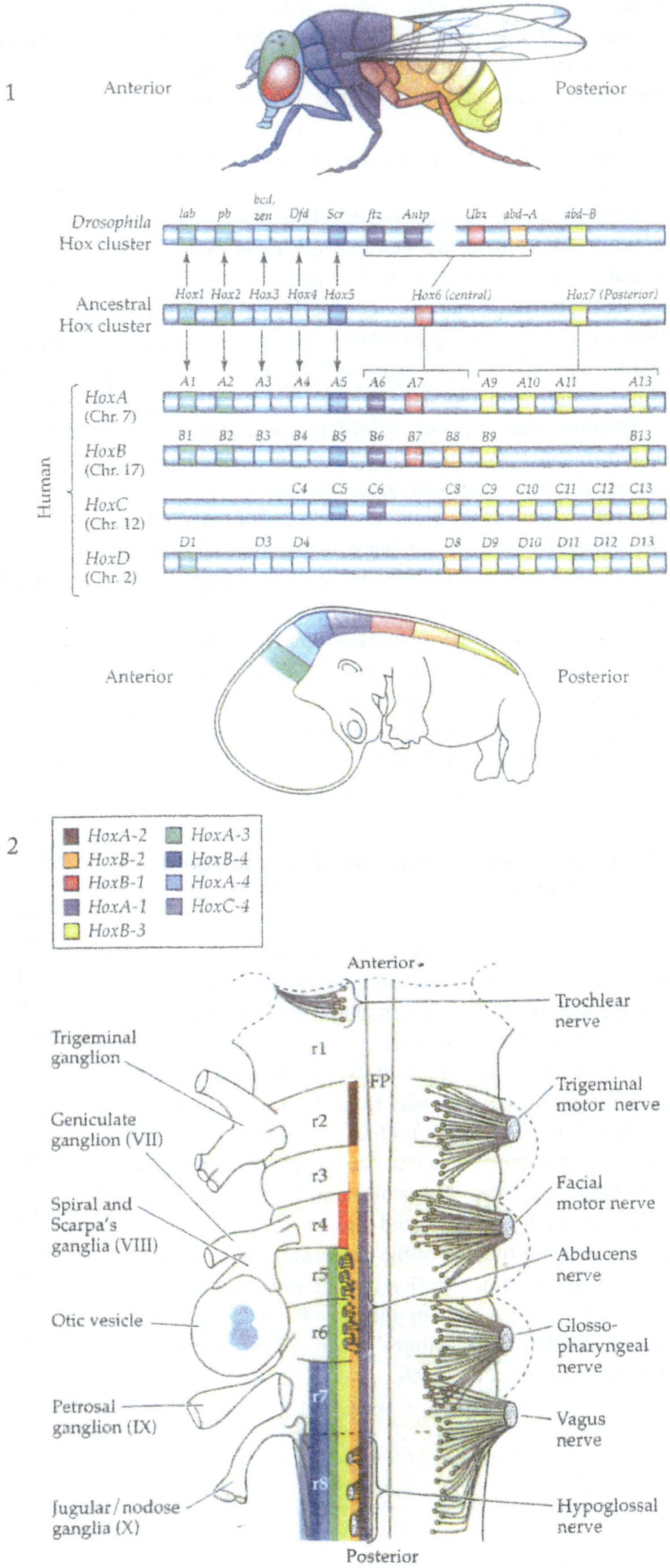

Fig. 19.3 Order in the brain is dictated by gene order at a previous level. (**1**) Human *Hox* genes (and those of most mammals) have apparently been duplicated twice, leading to four independent groups (*HoxA*, *HoxB*, *HoxC* and *HoxD*) each on a distinct human chromosome (7, 17, 12 and 2 respectively). The anterior to posterior pattern of *Hox* gene expression in both flies and mammals (including humans) follows the orientation of these genes in their respective chromosomes. (**2**) Parallel between early embryonic segmentation and early brain development. *Hox* genes *A*, *B* and *C* and related genes, are expressed in banded patterns in the developing chick nervous system, especially in the rhombencephalon which in the chick (and most mammals) consists of a series of seven transient bulges called rhombomeres. These seven segments are sites of differential cell proliferation, mobility and adhesion. The Figure shows rhombomeres in the developing chick rhombencephalon (hindbrain) and their relationship to the pattern of *Hox* gene expression (*colored bands*) and the differentiation of the cranial sensory ganglia and nerves (*left*) as well as cranial motor or mixed nerves (*right*)

19.13 Unexpectedly *Hox* Genes Do Not Participate in Part of the Midbrain and the Whole Forebrain

As mentioned earlier, the mammalian embryonic hindbrain is segmented into seven rhombomeres along the body axis.

Significant is that *r1* and *r2* rhombomeres in the posterior midbrain are a *Hox-free* area—they are devoid of *Hox* gene expression. This means that the largest portion of the midbrain and the whole of the forebrain are *Hox-free* zones (Liu and Xiao 2011).

The role of the neural crest in the construction of the vertebrate head was established in avian embryos. The entire facial skeleton and most of the skull were in part derived from rhombomeres *r1* and *r2*. To obtain further information of the *Hox-negative* domain, segments of the chick embryo were removed surgically, indicating that the neural crest is an important signalling center for brain development (Le Douarin et al. 2012).

Sources of Figures

Fig. 19.1 Friday, A. and Ingram, D.S. (Editors) 1985. The Cambridge Encyclopedia of Life Sciences. Cambridge University Press, Cambridge, UK (Fig. 2.26, page 90).

Fig. 19.2 (**1**) Finlay, B.L. and Darlington R.B. 1995. Linked regularities in the development and evolution of mammalian brains. Science 268: 1578–1584 (Fig. 1, page 1578). (**2**) Rilling, J.K. 2001. Allometric departures for the human brain provides insights into hominid brain evolution. In: Finlay, B.L. et al. 2001. Developmental Structure in Brain Evolution. Behavioral and Brain Sciences 24: 263–308 (Fig. 1, page 292). (**3**) Schoenemann, P.T. 2001. Brain scaling, behavioral ability, and human evolution. In: Finlay, B.L. et al. 2001. Developmental Structure in Brain Evolution. Behavioral and Brain Sciences 24: 263–308 (Fig. 1, page 293).

Fig. 19.3 (**1**) Purves, D. et al. (Editors) 2012. Neuroscience. Sinauer Associates, Sunderland, MA, U.S.A. (Fig. 22.4, page 483). (**2**) Purves, D. et al. (Editors) 2012. Neuroscience. Sinauer Associates, Sunderland, MA, U.S.A. (Fig. on page 484, Part *A*).

20 The Brain Turns Out to Have Its Own Geometry

20.1 The Brain Has an Internal Geometry

Any picture that we see has normally well-defined contours. A cube, or a cylinder, placed in front of us, can be recognized by their sharp edges and surfaces. They could be perceived by the brain as blurred images, but instead they are seen as defined by their straight lines and clear cut circles.

How could the external geometry of solids be apprehended by neurons if they had no architecture of their own, i.e. if they themselves did not follow geometry? This question would be considered out of context or irrelevant a few decades ago. But experimentation with animals and microscopic analysis of the cellular organization of the brain has progressed so rapidly in the last years that a concrete answer is now available. Unexpectedly the brain turns out to have its own geometry.

To start with, the primary visual cortex has distinct anatomical layers, easily recognized with the light microscope, each with characteristic input and output connections. Axons of neurons originating in the lateral geniculate nucleus extend to these layers terminating at specific sites. Some of these cells have excitatory, others inhibitory activities. Local information continues to flow to cortical, subcortical and other types of areas in an ordered way. The result is that each layer has different outputs.

20.2 Brain Cells Have Axes of Orientation: How to Get a Nobel Prize

Already in the late 1950s Hubel and Wiesel (1959, 1962) discovered that neurons had different excitatory and inhibitory zones. Three additional features were: (1) The cells of the retina and the lateral geniculate nucleus had a pattern of organization different from that of the "simple cells" of the cortex. (2) The receptive fields of excitation and inhibition were located following different geometric figures. (3) The shape of the cells was circular in the first group and elliptical, or rectangular, in the second cell group. On one hand, in the circular cells of the retina the receptive fields were also distributed circularly, the inhibitory fields being in the center of the cell and the excitatory on the periphery. The reverse situation also occurred but was accompanied by a comparable geometric distribution. These experiments were carried out on 24 cats using microelectrodes attached to the skull. Cells were classified as "simple" or "complex" depending on their behavior in relation to light stimulation building either simple or complex receptive fields in the visual cortex.

Additionally, in the neurons of the primary visual cortex a "simple" cell responded to a bar of light with an axis of orientation that corresponded to the axis of the cell's receptive field. Moreover, these cells had all axes of orientation: vertical, horizontal and oblique. The inhibitory and excitatory zones are distributed linearly within the cell building a central rectilinear axis or an external axis, or even quadrants of these fields, all following a most perfect geometry. For their "classical work that created a revolution in our understanding of the visual processing" Hubel and Wiesel became Nobel Laureates in 1981 (Robertson 2009).

20.3 Experiments with Monkeys Show Differential Response to Geometric Patterns: The Cells of the Brain Follow the Geometry of the Object Displayed

Experiments were also carried out with living monkeys. Among other information it was found that a cell in the primary visual cortex responded maximally to a bar of light that was parallel to the axis of the cell's respective field (Hubel and Wiesel 1979).

When the orientation or position of the bar of light was changed, the cells of the brain of the monkey, responded differently following the geometry of the object displayed. The cells responded optimally to a certain orientation of the stimulus. Light coming from a vertical edge of a square evoked a strong excitation wherever it was placed in the

A. Lima-de-Faria, *Molecular Origins of Brain and Body Geometry*,
DOI 10.1007/978-3-319-06056-9_20,

field. Illumination of the entire field of vision, with no edges, did not excite the cell.

20.4 Neuronal Cells Are Organized in Columns that Respond to a Single Axis of Orientation

The geometry also extends to the way the cells are distributed in the brain.

The cells with the same response properties are organized in columns. Some cells respond only to horizontally oriented stimuli and neurons within a column respond to a single axis of orientation. Each orientation column is about 30–10 microns wide. The interconnection in columns permits a new level of abstraction of visual information and is a first step in the deconstruction of the visual world.

The regularly oriented columns communicate with one another by means of horizontal connections between cells within a single layer. Cells separated by about one millimeter can fire simultaneously in response to stimuli with a specific orientation and direction of movement (Kandel et al. 1995).

Geometry turns out to be present at every level of brain organization. It is thus not surprising that we can conceive ideal circles and ideal solids in our brain. Our neurons seem not to be able to do anything else due to their internal geometry and their oriented connections.

21 Left and Right Side, of the Body and of the Brain, Are Not Mirror Images and Behave Independently

21.1 Independence of the Left and Right Side of the Body in Crabs, Snakes and Humans

A minor detail can unveil an important underlying phenomenon.

In most snakes the dark bands run across the cylindrical body, but in the Coral pipe snake the stripes are interrupted in three areas on the dorsal rim. There is a distinct shift in the location of the bands on the two sides of the body. They are out of phase, not building a mirror image. This reveals an independence between the left and right side of the body during the formation of the pattern (Fig. 21.1).

The Tiger salamander *Ambystoma tigrinum* has a black body covered by yellow blotches. These occur on both sides of the body in near equal numbers. As in the snake, they do not form mirror images. The left and right sides of the body were built independently.

Crabs are well-known for differences between the right and left halves of their bodies. The Sand fiddler crab (*Uca pugilator*) has a large claw that grows to impressive dimensions on the right side of the body. Another closely related species *Uca vocans* displays an equally large claw, which is brightly colored, but is located on the left side. Both cases disclose the independence of the two halves.

Most humans show external differences between the right and left side of the body, the best known is the use of the hands, some are left-handed others right-handed. If one faces a mirror and closely inspects one's face it becomes evident that especially the eyes have different shapes on the right and left side, and tend to point differently. One eye looks more in a straight direction whereas the other is less so, being focused on a general view. This is detected in many photographs and artists have long discovered the significance of this asymmetry in their self-portraits such as Rembrandt van Rijn (1606–1669) and Vincent van Gogh (1853–1890). Pablo Picasso (1881–1973) with his unique observation of the human condition painted it on the canvas in a clear cut way. A compound face consists of an eye looking forward and the other directed to a side view.

21.2 The Surge of Epigenetics: The Expression of Genes Is Dependent on Their Parental Origin and Changes During Development

Science, like any other human activity, is prone to follow given trends. It is natural that scientists are led to ask questions that spring from the ideology dominating at a given time.

In the early days of genetics (1900–1950) the main preoccupation was to show how genes were inherited, but above all, how they maintained their permanence. Mutation, when it occurred, although it implied a change in gene structure or function, was also a modification to another permanent state. A corner-stone was the stability of the gene.

As a result a vast number of works and books were published on human twins which emphasized their genetic similarity and identical behaviour. When disparities arose these were considered marginal or without significance. Only mutation and selection were considered the causes of variation.

Embryologists, who studied in detail the programmed unfolding of organisms from a single fertilized egg could not agree on such an interpretation. A divorce occurred between the two camps. In the meantime, embryologists coined the term epigenetics to emphasize that genes could change their function during the development of an egg into an adult organism.

The concept was dismissed or ignored by geneticists in general until molecular biology became the dominant discipline in the last decade. Suddenly epigenetics came into the limelight and the variation of gene function during morphogenesis has become the present preoccupation. Epigenetics deals essentially with the different states that genes may

A. Lima-de-Faria, *Molecular Origins of Brain and Body Geometry*,
DOI 10.1007/978-3-319-06056-9_21, © Springer International Publishing Switzerland 2014

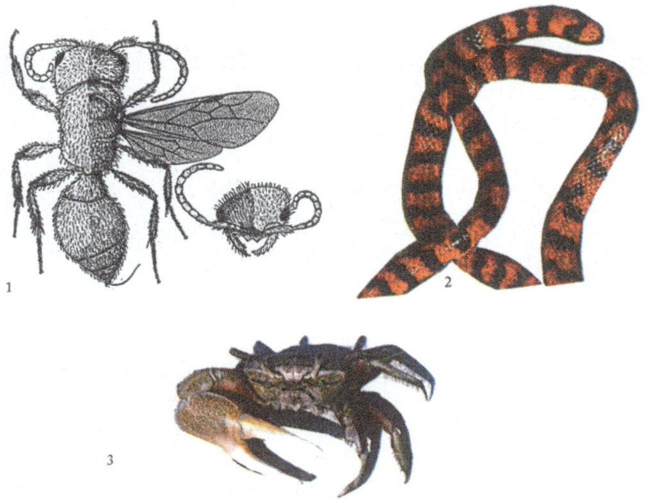

Fig. 21.1 The independence of the *left* from *right* side of the body in vertebrates and invertebrates. (**1**) The insect *Pseudomethoca canadensis*. (**2**) South American coral pipe snake *Anilius scytale*. Over 50 *black bands* appear along the body with an interval that tends to be of the same magnitude. In three regions, the bands on the *left side* of the body do not coincide with those of the *right side*, being out of phase, and not building a mirror image. (**3**) Sand fiddler crab *Uca pugilator* with its large claw on the *right side* of the body

assume during cell differentiation within the same organism. For instance, the expression of certain genes is dependent on their parental origin, i.e. they will be silenced depending on whether they are located on the chromosome set derived from the father or from the mother. Not less than 30 such genes have been identified in humans and mice (Feil and Khosla 1999). The epigenetic effect that leads to the imprinting of a gene has been found to be associated with methylation of its DNA. This means that other changes occur in DNA than its nucleotide sequence, which is considered the usual source of mutation.

These alternative states of gene activity, which are essentially epigenetic, elucidate the difference in development between the left and right halves of the body of humans and other vertebrates. It has been established that microRNAs play a role in the formation of epigenetic domains in humans (Grewal and Moazed 2003). These RNAs are also known to shape the road map of the genetic events directing cell differentiation in the developing embryo (Matzke and Matzke 2003).

Epigenetic gene changes are now considered to be also involved in disease. A growing number of cancer genes are being recognized that harbor dense methylation in DNA regions which normally are unmethylated (Baylin and Herman 2000).

21.3 A Dorsal Rim Band Separates the Left from the Right Side of the Body in All Groups of Vertebrates

As described above, in most groups of vertebrates—from mice to fish—there is a well delimited band that runs along the dorsal rim from head to tail (Fig. 6.1). It occurs in bright colours: black, red, yellow and white, dividing the right side of the body from the left side. This sharp band concretizes the independence of the two halves of the body by establishing a clear geographical frontier between them.

21.4 Cilia Discern Left from Right: The Molecular Mechanism Beyond the Emergence of Body Asymmetry in Mice

When the human body is observed externally it shows left-right mirror symmetry, that is accompanied by a minor but significant asymmetry between its two halves. Internally the picture is different, revealing a distinct asymmetrical distribution of several of its main organs, such as the heart or the liver.

In the mammalian early embryo, left-right symmetry is broken by the action of rotating cilia, which are small hair like protrusions on the surface of cells located in a pit-like structure called the *node*. Cilia are tubular extensions of the cell membrane containing microtubules associated with proteins. They are frequently sensory, and participate in motion, beating in a synchronized rhythm (Abercrombie et al. 1990).

In the embryo, the cilia generate a unidirectional flow of fluid across the node from right to left. This flow drives asymmetries in gene expression and Ca^{2+} signalling in cells at the periphery of the node. By studying development in mice Yoshiba et al. (2012) discovered a mechanism by which the embryo senses nodal flow (Fig. 21.2).

A series of genes and mutations resulting in defective genes were used to decipher the mechanism.

The asymmetric signals induced by flow are first apparent in the crown cells bearing immotile cilia. Here, the *Nodal* molecule is expressed more strongly on the left than on the right side. This molecule has an antagonist: the secreted molecule *Cerberus-like 2*. Subsequently *Nodal* targets the gene encoding the transcription factor *Paired-like homeodomain transcription factor 2*. Both are expressed more laterally.

Another component of the mechanism is the transmembrane cation channel *Polycystin-2* which is necessary to decide the left-right patterning of the mouse embryo. This

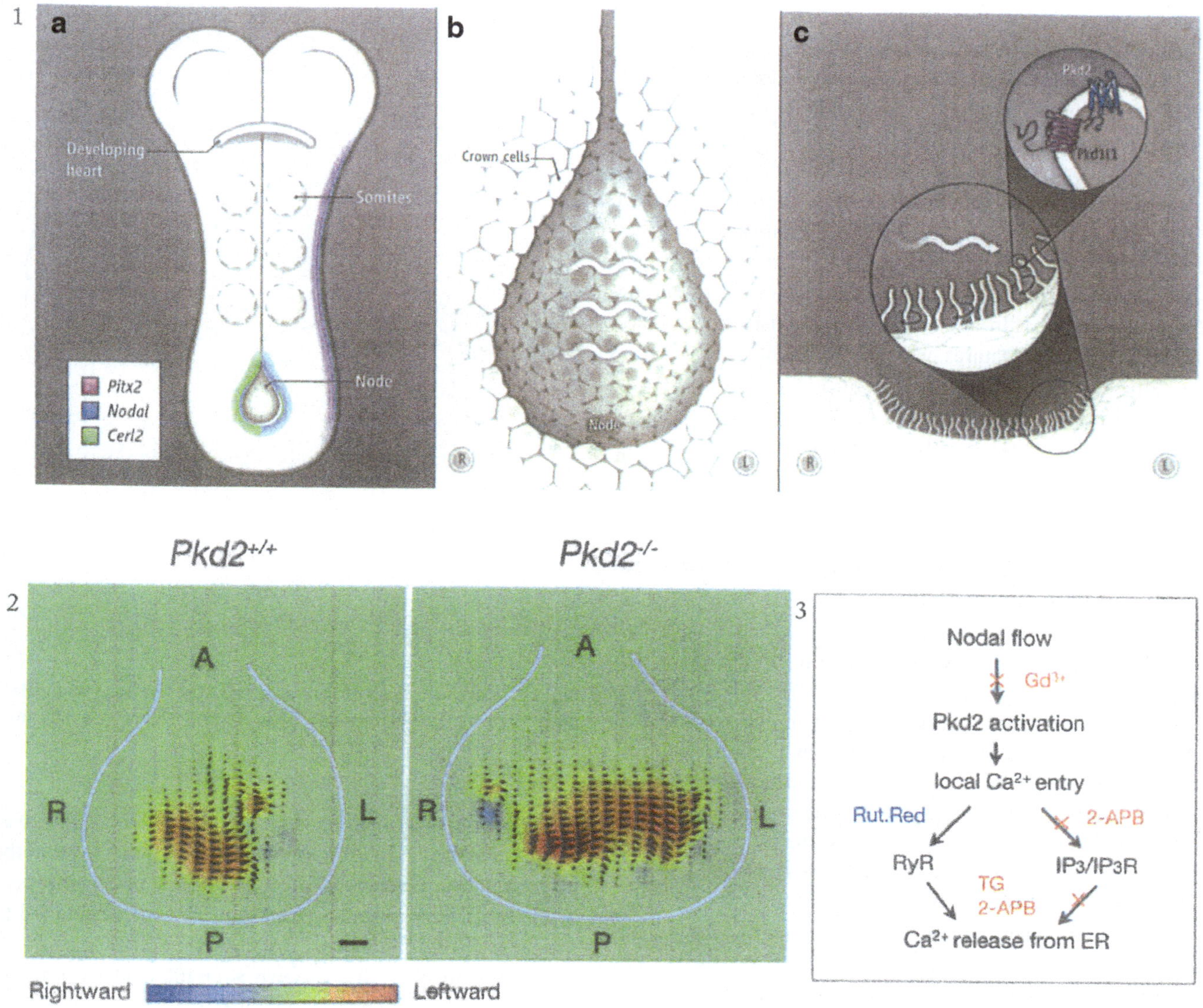

Fig. 21.2 ***Left–right*** **determination in mice.** **(1) Feeling the fluid flow.** (**A**) The early mouse embryo (day 8.25), showing asymmetric expression of the indicated genes around the node and laterally. (**B**) The node consists of a depression containing central pit cells with motile cilia and peripheral crown cells with immotile sensory cilia. (**C**) Cross-section through the node showing *Pkd1l1-Pkd2* complexes in crown cell cilia sensing leftward flow. The signalling pathway links fluid across the node with asymmetric gene expression. (**2**) **Mechanism by which the mouse embryo senses nodal flow in** ***left–right*** **determination.** The gene Pkd2 is necessary for sensing of nodal flow. Nodal flows in wild-type $Pkd2^{+/+}$ and the mutant $Pkd2^{-/-}$ embryos. Flow is maintained in the $Pkd2^{-/-}$ embryo. Each *arrowhead* denotes the direction and speed of the flow at the indicated position. The *color scale* indicates the direction and magnitude of the flow velocity (*leftward in yellow* and *red*, *rightward* in *blue*). *White lines* indicate the outlines of the node. Scale bar, 10 μm. (**3**) **Calcium signal is essential for L-R symmetry breaking in the node crown cells.** L>R asymmetric ANE activity is maintained in control and Ruthenium *Red*-treated embryos, whereas it is disrupted with Gd^{3+}, 2-aminoethoxydiphenyl borate (2-APB), or Thapsigargin (TG). Scale bar, 50 μm. *ANE* is an enhancer derived from the human *Lefty1* gene

is implicated in the elevation in intracellular calcium concentration (Ca^{2+}) that occurs to the left of the node in response to flow. Mutant mice exhibit several left-right defects in which these genes are involved and demonstrate a role of Ca^{2+} signalling in left-right asymmetry.

Application of exogenous rightward fluid flow resulted in activation on the right side of the normally left-restricted genes. The sensory proteins, together with their antagonists, are the main participants in the determination of the body asymmetry in mice (Norris and Grimes 2012).

21.5 Brain Asymmetry Is Widespread in Vertebrates

The lateralization of the brain was once thought to be an exclusively human attribute, associated with extraordinary language abilities (Gill 2003). Until the 1970s cerebral asymmetries were thought to be unique to humans, following the general tendency to consider *Homo sapiens* as a species apart, a reasoning that the study of its anatomy and functions has shown to be erroneous (Güntürkün 2012).

Cerebral asymmetry is a ubiquitous phenomenon. Pattern discrimination tasks often evoke a right hemisphere advantage in smaller land mammals like cats, rats and sheep. Rats have a tendency to first turn left when being placed into an open field. In aquatic mammals, such as dolphins, right-eye superiority implies left hemisphere dominance for visual object processes. Different species of birds display right-eye (left hemisphere) superiority during the discrimination of visual patterns. In humans the left hemisphere excels in identifying local features whereas the right hemisphere is usually faster and more accurate in identifying the global components of the input. This became apparent in brain-damaged patients. The evidence clearly points to a common heritage of cerebral asymmetries (Güntürkün 2012).

21.6 Brain Anatomy and Function Show Clear Cut Asymmetries

There has been some controversy concerning anatomical asymmetries which has now been dispelled (Eccles 1989).

No asymmetry has been observed in the monkey and baboon brains but in the human brain this is now well-established for adult brains as well as in new-born babies. The *Sylvian fissure* was found to be different on the right side from the left side. This was related to the asymmetry of the *planum temporale* which is larger on the left side in 65 % of the cases (Fig. 21.3).

The functional asymmetries take many forms. For monkeys the asymmetry is provided by hand preferences. During hominid evolution an enormous development of asymmetries in the functions of the hemispheres has taken place (Eccles 1989). These differences became obvious in patients treated by section of the *corpus callosum* as well as localized lesions. Notable is the asymmetry of speech areas. Large parts of the left parietal and temporal lobes are specialized in the semantics of speech recognition and production. The right parietal lobe is especially involved in the handling of spatial data and the relationship between body and space. Lesions result in loss of skills.

21.7 The Two Hemispheres Have Different Cognitive Capabilities but Complement Each Other

Patients who underwent brain operations in which the *corpus callosum* and the *anterior commisure* were sectioned, preventing the communication between the two brain hemispheres, showed that each hemisphere is capable of functioning independently. But the right hemisphere is generally mute and cannot communicate its experience verbally.

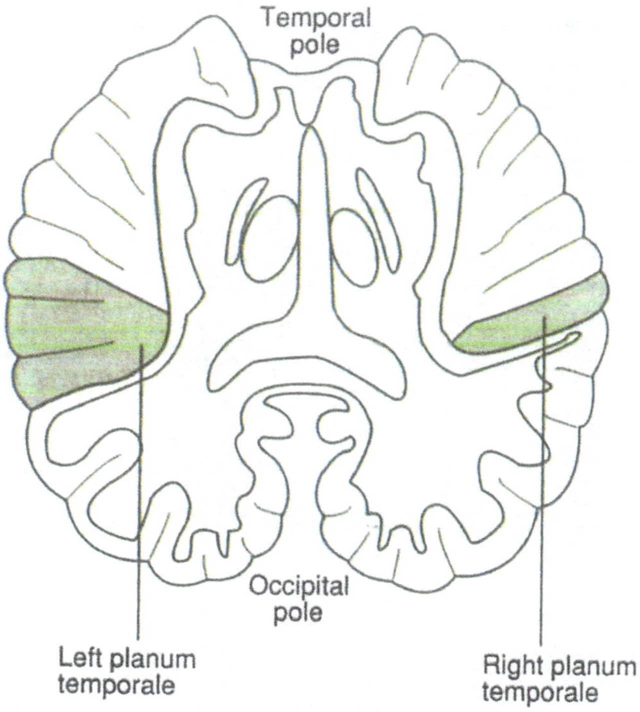

Fig. 21.3 In the majority of human brains the planum temporale is larger in the left hemisphere than in the right (Adapted from Geschwind and Levitsky 1968)

Experiments with animals in which the *optic chiasm* was also cut (destroying the cross visual fibers) indicated that the corpus callosum conveyed information from one hemisphere to the other during training. Subsequent experiments led to the same result in humans.

Each hemisphere has its own strengths and weaknesses with regard to a given task.

However, the left hemisphere excels in intellectual, rational, verbal and analytical thinking and the right hemisphere excels in sensory discrimination and in emotional, nonverbal and intuitive thinking. The two hemispheres seem to be necessary for harmonious behavior (Kandel et al. 1995).

21.8 Bird Song Is Normally Controlled by the Left Hemisphere

Bird singing is controlled through a distinct pathway from the brain to the sound production apparatus, the syrinx, located in the base of the tracheal windpipe. Nerve impulses start in the forebrain, proceed to the midbrain, and finally to the nerves of the syringeal muscles.

Bird calls can be evoked by electrical stimulation of the midbrain. Destruction of this center renders a songbird mute.

The song is normally controlled by the left hemisphere, this part of the brain being dominant in canaries as it is in humans (Gill 2003). The right cerebral hemisphere assumes control of the functions of the left hemisphere only if the left hemisphere is damaged.

Impairment of a young bird's song-control centered in the left hemisphere leads to the formation of an alternative set in the right hemisphere and the acquisition of a new song repertoire.

The song-control centers of canaries and finches are controlled by the sex hormones: *estradiol* and *testosterone*. These increase the length and branching complexities of dendrites—receptor branches—of some neurons in the song centers, and grow new or replace old synapses. New dendrite connections are formed in the spring when the canaries learn new songs. These neural connections disintegrate in the fall when the birds stop singing. Adult birds can form new neurons and replace old ones. A neural growth factor, a large protein molecule, has been isolated from such neurons named *neuroleukin* (Gill 2003).

Sources of Figures

Fig. 21.1 (**1**) Wheeler, H. 1940. The Miracle of Life. Odhams, London, UK. (**2**) WIKIMEDIA.org/WIKIPEDIA/commons. (**3**) WIKIMEDIA.org/WIKIPEDIA/commons.

Fig. 21.2 (**1**) Norris, D.P. and Grimes, D.T. 2012. Cilia discern left from right. Science 338: 206–207 (Fig. page 206). (**2**) Yoshiba, S. et al. 2012. Cilia at the node of mouse embryos sense fluid flow for left-right determination via *Pkd 2*. Science 338: 226–231 (Fig. 2, page 229). (**3**) Yoshiba, S. et al. 2012. Cilia at the node of mouse embryos sense fluid flow for left-right determination via *Pkd 2*. Science 338: 226–231 (Fig. 3, page 229).

Fig. 21.3 Kandel, E.R. et al. (Editors) 1995. Essentials of Neural Science and Behavior. Prentice Hall International, Inc., London, UK (Fig. 9.11, page 539). *From*: Geschwind, N. and Levitsky, W. 1968. Human brain: left-right asymmetries in temporal speech region. Science 161: 186–187.

22 How the Brain Changes the Body Geometry at Will

22.1 The Brain Decides What Color and Pattern Will Be Displayed: Chameleons Know How to Do It

Not less than 180 species and subspecies of chameleons have been described and their fossil record goes back to 26 million years but may even extend to 60 million years.

The chameleon is the classic example of an animal's ability to change its color rapidly according to environmental situations. The analysis of this phenomenon discloses many details which elucidate in a remarkable way the diversity of mechanisms involved in coloration.

To start with, its turretlike eyes can move independently of one another allowing the chameleon to have binocular vision and the ability to judge distance with great precision. It scans the surrounding environment with "phenomenal eyesight" enabling it to defend itself or attack prey (Halliday and Adler 2004).

When chameleons target prey, two separate images merge into one to gauge distance. Then the extendable tongue, which consists of cartilage, muscles, nerves, glands and tissue, is projected engulfing the prey in fractions of a second (Grzimek 2004).

Several strategies are employed when defense is necessary. The body has often a high arch on the back and a coloration that is close to that of the leaves that surround it. These two features give several species of chameleons a leaflike appearance. The effect is enhanced by an oscillation of the body, back and forth, like that of leaves in the breeze (Madagascar species).

The rapid change in coloration is the most powerful defense mechanism. It is important to realize that this signal is given by the eyes, with their phenomenal accuracy of vision, and further transmitted to the brain, which in turn sends a message to the cells that contain pigments—the chromatophores. These are located in the chameleon's skin formed by five layers: epidermis, basement membrane, xanthophores (red), iridophores (blue) and melanophores (brown). Each chromatophore has a central pigment saccule surrounded by muscular appendages. These contract or expand diminishing or increasing the color of the skin. The action does not occur throughout the whole body but at specific sites. The result is that not only the color is changed but the pattern is also altered following decisions emanating from the brain.

The analysis of most chameleon species reveals the following: (1) The palette of any species is limited to certain colors. (2) Members of the genera *Brookesia* and *Rhampholeon* are mainly brown and black. (3) Males of *Furcifer pardalis* (panther chameleon) have distinctive color hues such as pink combined with blue, green with red, aqua blue with green, red-orange with white and turquoise with navy blue.

It was earlier assumed that these colors resulted from an adaptation of these reptiles to the surrounding environment. This idea has been supplanted in face of more detailed studies. As Grzimek (2004) writes: "Chameleons display variations of their color palettes in response to psychological or physiological stimuli and to communicate, not to match their background as once believed". In some species they may resemble leaves in color and shape, but this is only one aspect of their coloration behavior (Fig. 22.1).

The mental decisions take many forms: (1) A male communicating its intention to mate displays vivid colors. A female likewise will show her willingness by subduing her color, but if she is not receptive she will present a dark intense hue or purple spots. (2) During competition between males the losers change to drab coloration. (3) During excited or stress conditions chameleons modify their colors on different parts of their body. These take the form of stripes and bands that recede when the chameleon becomes calm.

Independently of the variable pattern, the chameleon has an underlying ground coloration and design that is constant and which is characteristic of each species. The female of *Furcifer minor* has large yellow bands, parallel to one another, set against a dark green background. They are evident on the head, trunk, tail and legs. Of interest is that

A. Lima-de-Faria, *Molecular Origins of Brain and Body Geometry*,
DOI 10.1007/978-3-319-06056-9_22, © Springer International Publishing Switzerland 2014

1 2

Fig. 22.1 The rapid change of *coloration* in chameleons results from three types of chromatophores and their expression or repression. (1) Three color patterns in Smith's dwarf chameleon *Bradypodion taeniabronchum* used in different situations connected with sex relations and thermoregulation. **(2) Change of pattern in the same individual within 24 h.** The Cichlid fish *Aequidens tetramerus*. Four of the seven changing patterns observed within 24 h in a single living fish. Parallel longitudinal stripes, which run along most of the body, are transformed into parallel transversal stripes. At the same time, the *circular black spot*, in the middle of the body, does not change either position or shape

their orientation in relation to the body parts follows the main orientation seen in zebras. In this mammal, as in this reptile, the bands on the legs have a horizontal orientation, whereas those of the head, trunk and tail tend to have a vertical position. Another quite closely related species of chameleon, since it belongs to the same genus (*Furcifer campani*), has instead three longitudinal bands that traverse the whole body. Again, this reptile has a pattern with the same band orientation and distribution as the mammal *Galidictis grandidieri*.

The chameleons reveal two separate processes. The first is a permanent pattern that is characteristic of the species. The second lies in the ability of the brain to direct the emergence of a novel pattern and color that last for a period that the animal considers convenient.

22.2 Pulsating Stripes in Cuttlefish, Used in Tactical Deception, that Appear and Disappear in Only Half of the Body

In mollusks the predominant function of the brain, in deciding the pattern displayed by the organism, could hardly be more evident.

Cuttlefish are cephalopods known for facultatively changing their shape and color. Males of *Sepia plangon* (mourning cuttlefish) exhibit a pattern of pulsating stripes. During courtship they deceive their rivals by displaying a specific male pattern on the side of the body facing a receptive female and simultaneously displaying a female pattern on the opposite side that faces a rival male. This strategy prevents the rival from disrupting courtship (Brown et al. 2012) (Fig. 22.2).

Other cephalopods, such as the Caribbean reef squid, are known to use laterally split body pattern as a form of defense against predation. These observations were carried out in the wild by diving and photographing during the breeding season. They were accompanied by aquarium studies.

Two phenomena become evident. First, a mental decision that determines the rapid change of pattern by the use of pulsating stripes. Second, the distinction between the two halves of the body due to the ability to separate the left side from the right one.

The use of tactical deception, a complex communication process, is a mark of the high cognitive capacity of cephalopods. It is well known that the brain in these invertebrates is more comparable to that of vertebrates than of other molluscs. Their highly complex behavior includes

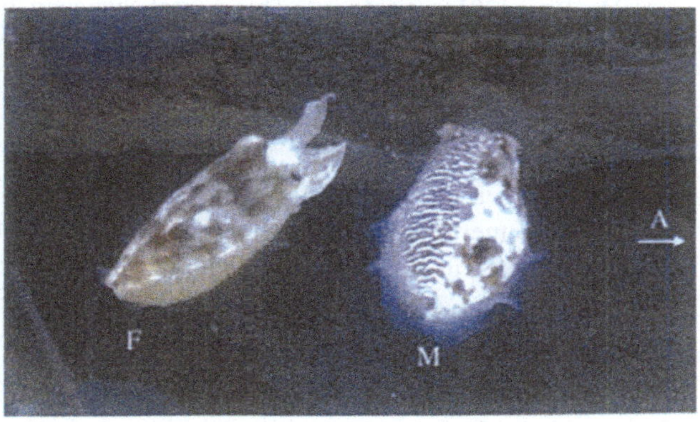

Fig. 22.2 **Tactical deception in a cephalopod by pattern change—the *left side* of body shows a pattern different from the *right side*.** Male mourning cuttlefish (**M**) displaying a male-specific pattern towards a female (**F**) while simultaneously displaying deceptive female *coloration* towards a rival male (**A**)

tool use. Their cognitive capacity has been considered to parallel that of primates.

22.3 Dynamic Imitation of Other Species by an Octopus

Camouflage and the ability to imitate other species are well documented in cephalopods.

Recently Norman et al. (2001) have described a more advanced form of behavior in this animal group. They found a species of octopus in Indonesia that is able to switch back and forth between mimicry of different model organisms. Individuals displayed a repertoire of postures and body patterns, several of which "are clearly impersonations of venomous animals" co-occurring in their neighbourhood. They imitated flat-fish as well as lion-fish (*Pterois sp.*) and sea-snakes (*Laticauda sp.*). The authors concluded that the octopus makes decisions about the most appropriate form of dynamic mimicry to avoid threats.

22.4 Release of Hormones that Determine Color and Pattern Changes in Flatfish

Biologists call flatfish "the chameleons of the fish world". A long series of experiments have been made with flounders and soles disclosing their change in both color and pattern depending on the surrounding environment. The result is that they resemble their background.

In an initial experiment, the flatfish *Paralichthys albiguttus* was kept for 3 h lying over a background that was half white and half black, the two fields of color being delimited by a sharp line. The head of the fish was maintained on the white side and the animal was immobilized. After this period, the fish had turned its body color to white, an indication that the information that led to the color change was transmitted by the eyes (Cuénot 1932).

This was confirmed by finding that blinded flat fish were not able to modify their color when moved to a different environment (Veil 1938).

Subsequently the mechanism became elucidated. The wavelength of light received by the eyes resulted in release of adrenalin and other hormones that modified the color of the animal by altering the shape of the pigments of the body (Eckert and Randall 1978).

The color changes were not based on changes in cell shape which assumed a relatively fixed form. Instead the response to hormonal stimulus led their pigment granules to spread or to concentrate centrally within the cell (Kardong 1995).

Hence the information received through the eyes, was carried to the brain. This in turn directed the release of hormones that targeted the pigment on skin cells modifying their color and pattern within a few hours (Withers 1992).

22.5 Adaptive Camouflage in Fishes Takes Place in Seconds: The Extreme Speed of Brain Action in Changing Body Geometry

To determine the rapidity of the change in color and pattern in fishes Ramachandran et al. (1996) performed experiments using the tropical flatfish *Bothus ocellatus*. The animals were housed in a saltwater aquarium and transferred to tanks with the bottoms having different backgrounds: coarse gravel, printed checkerboards, homogeneous grey and yellow sand. In the home tank all fish appeared almost identical in color and pattern. When transferred to different bottom backgrounds each changed its surface markings within 2–8 s, resembling closely the new background. Each fish was again removed to another tank, at random, and again it adapted rapidly to the new environment.

This species of flatfish happens to resemble chameleons in another feature. Their eyes are mounted on short stalks and move independently through 180°.

22.6 Longitudinal Stripes Are Converted to Transversal Bands Within Hours

Zebras are one of the best examples of a permanent pattern that is not only transmitted to the progeny in intact form but which is also fairly constant from juvenile to adult life. Their body pattern is mainly characterized by parallel vertical stripes covering most of the neck, trunk and tail combined with horizontal stripes along the four legs.

Patterns, such as this, seem fixed by the genetic constitution of the organism and as such immutable in principle or only modifiable by mutations. But the brain is much more powerful than we tend to believe. It can actually radically change the orientation of body stripes by 90 degrees and it does so within hours. This is actually what happens in the cichlid fish *Aequidens tetramerus* which may assume different forms within this space of time. Seven parallel stripes run along most of its body, soon they are reduced to a single horizontal stripe and at a later stage four vertical stripes begin to appear near the tail and soon nine vertical stripes cover most of the body and head. The whole event takes 24 h (Fig. 22.1).

However, parallel to this impressive transformation there is a permanent component. The distinct black spot that is seen on the central region of the body was maintained and did not change position.

Two separate processes are at work: one that maintains the unaltered spot and another that dramatically modifies the stripe configuration. This last process, like in the flatfish, is related to light reception by the eyes, because it follows day and night light intensity—the horizontal stripes are displayed by day and the vertical by night (Beebe 1943).

22.7 Mice Behave Like Flatfish: They Change Their Color According to the Dominating Color of the Environment. Their Gene for Pigmentation Has Been DNA Sequenced

Rodents are known to have the largest variation in coat color among mammals. Deer mice (*Peromyscus polionotus* of North America) have been studied for almost a century because they showed a strong correlation between pelage color and soil color. Besides, the different coat colors were found to be inherited when crosses were performed in the laboratory. In habitats with dark loamy soils (with mainland vegetation) the mice had dark brown dorsal coats. Those that colonized beach areas, with sand dunes, had a lighter color which blended with the sand. This change is considered to have occurred recently because they reside on barrier islands formed along the coast (Hoekstra et al. 2006; Hoekstra 2011).

This is a situation comparable to that of flatfish which become dark in a dark background and light-colored on a white bottom. It is expected that in the deer mice the information is also received by the eyes and transmitted to the brain but no experiments have been carried out in which the animals were blinded as in the case with the fish.

Instead a recent genetic analysis revealed that one region of the genome contained a known pigmentation gene coding for *melanocortin-1 receptor* (*Mc1r*). The gene, with about 1,000 DNA base pairs, was sequenced in beach and mainland mice. Only a single mutation was found consisting of a nucleotide that differed between the mainland and the beach population causing in turn an amino acid mutation in the protein (arginine to cysteine).

As Hoekstra (2011) points out "This radical change in the *Mc1r* protein's chemistry is predicted to affect the structure of the protein and hence its activity". *Mc1r* acts as a switch in determining the type of pigment a cell produces. When turned on by an activator the mouse coat gets dark (eumelanin), when turned off, by binding to an antagonist, the light color emerges (phaeomelanin). The activity is controlled by the protein's amino acid configuration. The dark mainland *Mc1r* protein increases the activity leading to dark coloration, by contrast the *Mc1r* protein from beach mice produces light pigments.

22.8 The General Occurrence of the Gene Involved in Pigmentation from Lizards to Mammoths

Mutations in the *Mc1r* gene also cause changes in a wide diversity of vertebrates. DNA extracted from 14,000-year-old mammoth, excavated from the permafrost of Siberia, had the same mutation found in deer mice and the amino acid change in the resulting protein was also from arginine to cysteine.

New Mexico lizards that inhabit a locality called White Sands are light colored relative to others at different locations. These blanched lizards have a unique mutation that affects *Mc1r* signaling, like the mutations in beach mice. Snow geese also have a mutation in *Mc1r* that exclusively produces the dark coloration chacteristic of adult birds (Hoekstra 2011).

The same gene produces a similar final coloration that is expressed on the fur (mice) on the skin (lizards) or on feathers (birds).

22.9 Besides Light Other Physical Agents, Such As Temperature, Decide Coloration

Temperature has the same power as light. Experiments with Himalayan rabbits disclosed a distribution of pigment synthesis in the skin directly related to temperature. Ears, nose, extremities of legs, and tail became black, whereas the rest of the body remained white. Temperatures (in degrees centigrade) that had a threshold of 1–2 degrees led to white, whereas 14–29 degrees produced black areas. If the black hair was cut and the temperature lowered, white hair was found to grow in the same spot, and if it was cut again and the temperature raised, hair then grew dark once more. In all cases a higher temperature resulted in darker color (Schmalhausen 1949) (Fig. 22.3).

It is well known that the main pigment of animal hair is melanin and that a change in hair color results from the combination of melanin with oxygen atoms leading to a

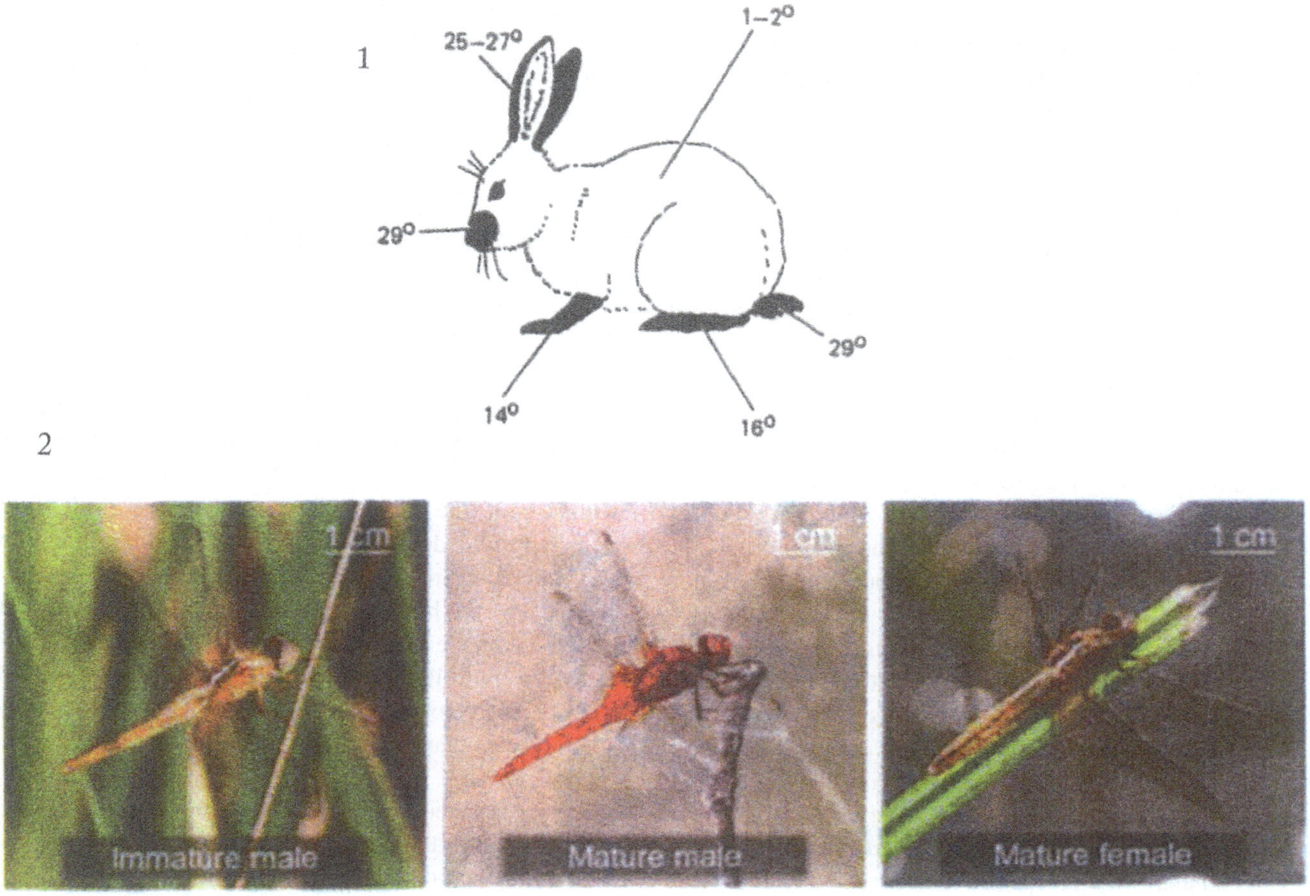

Fig. 22.3 (1) **Himalayan rabbit. The distribution of temperature thresholds for pigment synthesis in the skin (in degrees centigrade).** High temperature results in *black hair* whereas 1–2° leads to *white hair*. (2) ***Color* change in dragonflies due to simple atomic processes.** *Crocothemis servilia*. (*Left*) Immature male that is *yellowish*. (*Center*) Mature male fully *red*. (*Right*) Mature female that is also *yellowish*. The pigmentation changes are due to the role of oxidants and reductants as experimentally demonstrated

different chemical configuration of the pigment (Atkins 1987). It is important to note that the rabbit could not produce green, blue or red hair, it only changed from white to black and vice versa. There is plasticity but within well-defined limits. The color transformation in the rabbits involved oxygen. Another color change has recently been experimentally demonstrated in insects that also involves this atom.

22.10 Simple Atomic Mechanisms, Like Oxidation-Reduction States, Decide Yellow/Red Color in Dragonflies

Methylene blue is colorless when reduced, blue when oxidized—the classical example of a substance which undergoes a definite color change during its reversible oxidation or reduction.

Such an atomic mechanism alters yellow dragonflies into red. In these insects body color change is associated with sexual maturation as it is in many vertebrates. In some dragonflies, whereas females and young males are yellowish, aged males turn vivid red upon sexual maturation. The yellow/red color changes in the dragonflies are regulated by oxidation-reduction states of epidermal pigments. This was demonstrated by injecting the reductant solution into live insects, the yellow-to-red color change was experimentally reproduced in vivo in immature males and mature females which changed their color to red within hours. On the contrary by injecting an oxidant solution into the abdomen of mature males they turned from red to yellow.

Following laboratory experiments the extracted pigments changed color according to the oxidation-reduction conditions. "An impressively simple molecular mechanism" is responsible for this radical change in the coloration of living organisms (Futahashi et al. 2012) (Fig. 22.3).

22.11 Body Coloration and Pattern Are Radically Modified During Animal Development

Most colorations and patterns are strongly determined by the genetic make-up of the organism. They are evident at birth and are maintained until late adulthood. The examples are many from mice to humans.

But this is not a general rule among vertebrates. Instead, there are well-known examples among mammals, fishes and birds, that radically change their coloration and pattern throughout life. Young tapirs have a red-brown coat with horizontal light stripes and spots. Similarly, piglets of wild boar are covered by long horizontal brown bands separated by light colored ones. In both cases the adults get rid of this pattern, becoming totally dark grey from nose to tail. The adult coloration begins to appear at about 6 months (Burnie 2004).

The emperor angelfish *Pomacanthus imperator* did not get its name by accident. As an adult it is a most beautiful animal covered by over 20 deep blue stripes alternated by clear yellow bands. These stripes have mainly a horizontal position but the same animal, as a juvenile, had a large white circle located close to its caudal fin that was surrounded by less circular stripes which became vertical near the head. Development in this species transforms circularity and vertical stripes into horizontal ones (Fig. 22.4).

The same phenomenon is well-known in many bird species which are born with a rather even and drab coloration that, in the adult, is transformed into a most vivid and diverse rainbow (Harrison 1975).

The brain of fishes as well as that of other vertebrates, develops during the growth from the juvenile to the adult stage, changing in structure and function during this period of time.

22.12 The Human Brain Also Changes Between Juvenile and Adult Stages

At birth, in humans, the brain had been considered to assume its final form and no new cells were thought to be added during later growth. This assumption has been superseded by recent research.

By the 10th week of pregnancy nerve cells start to be built at an explosive speed—about 200,000 per minute. Following birth, the human brain increases appreciably in weight. At birth it is 350–400 g whereas the adult brain is 1.4 kg (about 10 % and 2 % of the body weight respectively). The whole brain "awakes" as one is born. *Immediate early genes* like: *c-fos*, *c-jun* and *nurr-1* are activated. The proteins built from these genes are transcription factors which regulate other activities (Lagercrantz and Olson 2007).

It is difficult to understand how the approximately 20,000 genes identified in humans can participate in the organization of 100 billion neurons connected by 1,000–10,000 synapses. This number becomes even larger as the glial cells are added. These are more numerous than neurons by a ratio of perhaps 3–1. Based on current estimates, of the 20,000 genes, only some 14,000 are expressed in the developing and mature brain. Of this subset, about 8,000 are expressed as well in cells and tissues. A great deal of the brain-specific genetic information is considered to reside in introns and other regulating DNA sequences controlling timing, quantity, variability and cellular specificity of gene expression (Purves et al. 2012). This indicates that the brain has an enormous capacity for self organization.

The vigorous building of neurons that occurs before birth slows down after this event. For a long time it was believed that no new neurons were formed. At present it has been established that new nerve cells are formed in the brain late in life at two regions: the hippocampus (an area of synaptic plasticity) and the olfactory epithelium from the lining of the nose. These are the only neurons that continue to divide and differentiate throughout the organism's life (Lackie and Dow 1995). Besides, novel synapses, between nerve cells, are created during the whole life span at the impressive speed of 1 million per second.

During and after puberty there is an increase in myelin (a medullary sheath). This process called myelinisation enormously increases the speed of nerve communication. The whole brain continues to reshape itself and improves its internal connections throughout the whole adult life (Lagercrantz and Olson 2007).

Much remains to be investigated before we know what kind of changes in the brain participate in the shaping of the outer structure of vertebrates.

Fig. 22.4 Comparison between juveniles and adults in mammals, fishes and birds. (**1**) Piglets with *horizontal striped coats* with their wild boar mother (*Sus scrofa*). (**2**) Adult emperor angelfish *Pomacanthus imperator* and juvenile emperor angelfish. A pattern consisting mainly of *concentric circles* is transformed into over 20 *horizontal stripes* which are parallel to each other along most of the body length. (**3**) The male Emu (*Dromaius novaehollandiae*) stays with his chicks for 5–7 months, by which stage the young birds are able to defend themselves. Note the similarity between the geometry of the Emu and that of the wild boar. In both juveniles the stripes are longitudinal and are *brown* alternating with *white*. The adults have in both cases an even *brown color*. This indicates that the same molecular cascade occurs in the bird and the mammal

Sources of Figures

Fig. 22.1 (**1**) Stevens, M. and Merilaita, S. 2011. Animal Camouflage. Mechanisms and Function. Cambridge University Press, Cambridge, UK (Fig. 13.1, page 238). (**2**) Gregory, W.K. 1974. Evolution Emerging. Arno Press, New York, U.S.A. (Vol. 2, Fig. 9.87, page 249). From Beebe, W. 1943. Pattern and color in the cichlid fish, *Aequidens tetramerus*. Zoologica, New York Zoological Society, Vol. 28: 13–16 (Plate 1).

Fig. 22.2 Brown, C. et al. 2012. It pays to cheat: tactical deception in a cephalopod social signalling system. Biology Letters. The Royal Society 2012: 1–4 (Fig. 1, page 2).

Fig. 22.3 (**1**) After Schmalhausen, I.I. 1949. Factors of Evolution. The Theory of Stabilizing Selection. The Blakiston Company, Philadelphia, U.S.A. From Pritchard, D.J. 1986. Foundations of Developmental Genetics. Taylor and Francis, London, UK (Fig. page 63). (**2**) Futahashi, R. et al. 2012. Redox alters yellow dragon flies. Proc. Nat. Acad. Sci. U.S.A. 109: 12626–12631 (Fig. 1, page 12627).

Fig. 22.4 (**1**) Wilson, D.E. and Mittermeier, R.A. (Editors) 2011. Handbook of the Mammals of the World, Vol. 2. Lynx Edicions, Barcelona, Spain (Fig. photo page 264). (**2**) Smith, M.M. and Heemstra, P.C. 1986. Smiths' Sea Fishes. Springer, Berlin (Fig. Plate 73, Nr. 204.9B and 204.9 (A)). (**3**) del Hoyo, J. et al. (Editors) 1992. Handbook of the Birds of the World, Vol. 1. Lynx Edicions, Barcelona, Spain (Vol. 1, Fig. page 102, photograph).

Part VI

The Brain Is an Unreliable Magician

23 The Concept of Soul Became Evident Following the Agricultural Revolution 8000 B.C.: How It Changed Since Then

23.1 The African Origins of Humans Date Back 8 Million Years: As Migration Led North, Food Became Less Abundant Demanding Novel Solutions

Archeologists and anthropologists have established a main trend of human migration.

The earliest humans are considered to have appeared in Africa (which is still inhabited by chimpanzees and gorillas). The skeletons revealed that early humans were small and had differences in the pelvis, spine, knees, hips and toes, which made it possible for them to walk upright. Human origins go back 8 million years and the first hominid species, *Australopithecus*, identified in Ethiopia, lived 5 million years ago.

At that time humans lacked both tools and fire. By 2 million B.C., *Homo habilis* emerged who was able to produce tools which were found alongside the remains of butchered animals.

As ice-sheets, which had covered a large part of the earth receded, humans left Africa, where food was plentiful in the tropical rainforest. Juicy fruits and tender leaves could be collected everywhere among a bountiful and inviting nature. Chimpanzees and gorillas still enjoy this type of life.

By 1 million years ago our present species, *Homo sapiens* started travelling north. Following the last Ice Age (20000 B.C.) temperatures went up and much more land became available for human colonization. Soon, vast areas were covered with deciduous trees and pine forests. Humans were confronted with a different landscape, much less inviting than the tropical forest and which was partly hostile (Parker 1986; Hattstein 2011).

23.2 Archeological Research and Dating Techniques Established that the Agricultural Revolution Occurred 8000 B.C.

Leaves of pines were not easily digestible and acorns of oaks were not as appetizing and aromatic as the tropical fruits that they had left behind. A nomadic life became obligatory, searching continuously for food that was difficult to obtain in the form of regular supplies. This situation lasted until they reached the great deltas of the Nile and the Ganges. Here, there was a novel opportunity. Seeds, collected from shrubs and trees, could be pressed easily into the muddy soil of these humid plains and could be watched growing, giving rise to plants in a way that was not known before.

Humanity was witnessing one of its first scientific discoveries. Nature could be controlled in a way that obliged plants to produce regularly their progeny at a specific time and location. The enslavement of plants had become a reality. Dating techniques led to determine the first experiments, in what today we call agriculture, to 10000 B.C. By 8000 B.C., this technology had become a general phenomenon in localities where the soil conditions and climate allowed an easy plant cultivation.

Humanity's most profound revolution had taken place. After it human society assumed its basic form and would not change its organization for the following millennia.

23.3 The Enslavement of Plants Was Followed by that of Animals and of Humans

Subsequently full-scale agriculture developed in China, India, the Middle East, Egypt and Mesoamerica. It was a simultaneous phenomenon as disclosed by archeological research. The next experimental step was the enslavement of animals. Once plants could be grown in abundance, animals trapped in the wild, could be encircled by fences and fed with the leaves and stems not used for human consumption. Animal domestication became an obligatory component of human activity.

An unexpected consequence was that mass production of plants and animals led to the emergence of professions. Plants and animals had to be collected and cared for at specific times and the management of the soil and the study of the weather conditions became imperative. Humans

A. Lima-de-Faria, *Molecular Origins of Brain and Body Geometry*,
DOI 10.1007/978-3-319-06056-9_23, © Springer International Publishing Switzerland 2014

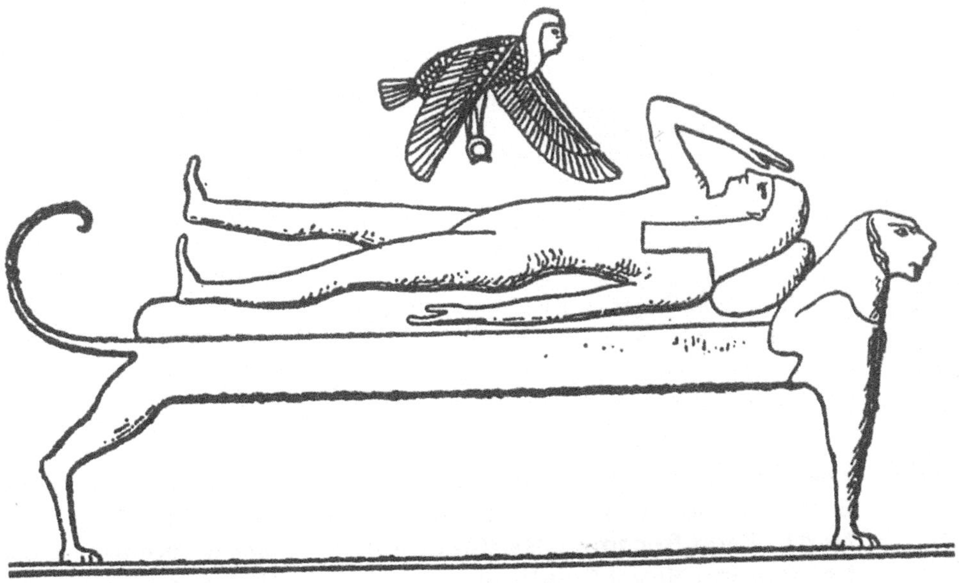

Fig. 23.1 **Representation of the soul** in the form of a bird-human chimaera that leaves the body but is material (Egyptian painting)

could no longer associate as loose groups of individuals, and societies with a hierarchical structure came into being.

It now payed to enslave other humans to be able to collect in a profitable way the vast amount of grain and cattle. Human slavery was established, to persist until the present, in different veiled forms.

23.4 The Search for Immortality Led to the Invention of the Soul

The number of professions and the society's hierarchy increased steadily. Between 7000 and 5000 B.C. a rigid social stratification became evident in early Egyptian civilization with the establishment of kings, priests, different types of functionaries and workers (Harris 2003). This was the society that later built the pyramids, which were not monuments but actually royal tombs. Their main purpose was the preservation of the bodies of kings assuring as much as possible their immortality.

It may seem too simplistic but the concept of immortality had its roots in two main physico–chemical events: death and birth. In a now stable society, they could be analysed in detail by studying the successive stages of life's origin and decay. Birth and death became such a preoccupation that they were extended to the heavens. For the ancient Egyptians the sun was born each day, as it rose on the horizon, and died each day, as it set below the earth's line.

At the same time another phenomenon seemed to go for ever. The stars and planets repeated their movements across the sky in a way that seemed to be independent of time. Immortality was the logical conclusion, but since the body, even that of kings, could not be saved from death and decomposition, it became mandatory to establish two separate independent entities. One the body, the other had to be immortal. It was a critical moment—the soul that left the body was invented. It was the only rational way to combine the permanence of the stars with the dynamic cycle of birth and death that pervaded terrestrial and non-terrestrial events (Fig. 23.1).

To this day the separation between body and soul still remains a part of our thinking since we are dealing with a society with the same basic organization (Fig. 23.2).

23.5 Originally the Soul Was As Material As the Body

At the dawn of Greek thought consciousness was defined following mainly African thinking. Originally the soul was conceived as material, as air or fire, or as palpable matter. There was then no great difficulty in the interaction of a material mind with a material body and they could easily separate from each other since they were two constituents of equal value (Eccles 1989).

Soon, Greek society moved from a pastoral stage, being a simple coalition of small trading harbors, to the formation of large cities relying on commerce, and war.

The body-soul problem became a source of permanent debate. Slaves had to be different from the ruling class. Animals also had to be different from humans. Equally, the soul had to be different from the body, it had to be non-material.

Fig. 23.2 Soul escaping from the mouth of the dead in a form identical to the body. The soul is caught between an angel and a demon, both striving to possess it. Medieval artists represented the soul as a smaller human body coming out of a larger dead body (Stained glass from Bourges cathedral, France)

23.6 The Abstract World of Pythagoras Influenced Future Thinkers: "The Essence of All Things Is Number"

Pythagoras of Samos (an island in the eastern Mediterranean) who lived between circa 560 and 480 B.C. was a mathematician and astronomer. He is not the founder of geometry. This was developed long before by the Egyptians and the Sudanese, far away in Africa 2,000 years before he was born.

What is central with Pythagoras was his recognition of the great importance of numbers and of geometry not only to build tombs but to understand the material world surrounding us. His theory of numbers led him to rediscover the theorem on right-angled triangles named after him. The Earth was also considered to be a sphere, and not a flat structure, having the shape of an ideal geometric solid.

For Pythagoras the universe was all geometry. This is why he stated: "The essence of all things is number" (Ian et al. 2002).

His abstract thinking had great influence on Plato (429–347 B.C.), Euclid (circa 300 B.C.) and Descartes (1596–1650).

23.7 In Plato's Aristocratic View of the World the Soul Became Immortal

Plato of Athens founded the Academy, a philosophical school. For him knowledge becomes an ever more ideal process. In his book the "Republic", he states that knowledge: "is only to be achieved by an intellectual gifted élite, who have spent many years in unremittingly abstract intellectual activities" (Hornblower and Spawforth 1999).

An aristocratic and abstract view of nature dominated at that time in Greece. Plato himself considered aristocracy the ideal form of social constitution. For him soul and body were separate entities, and souls were immortal and endlessly reborn into different bodies.

It is thus not surprising that during the Middle Ages Plato was revered and to the present day his view continues to be accepted by several social groups.

23.8 Aristotle Already Opposed Plato's View: Direct Observation and Experimentation Should Be the Basis of Knowledge

The father of Aristotle (384–322 B.C.) was court physician to the king of Macedonia. At that time body dissection was practiced by doctors and taught to their sons. This allowed a direct knowledge of the anatomy of living organisms initiating experimental science.

In Athens he entered Plato's Academy. He expresses deep respect for Plato's contribution even when he is most critical of his way of thinking (Hornblower and Spawforth 1999). After Plato's death he left the Academy and traveled to Mytilene where he did the biological research on which his scientific writings are based. His observations, especially in marine biology, were unprecedented in their detail and accuracy. It took nearly 2,000 years (until the 1600s) for scientific research to reach again the same standard of accuracy. Later he amassed a considerable library and encouraged his students to undertake research projects in natural science.

Aristotle was the first Greek thinker to employ the validity of scientific judgement and to reject fallacies and paradoxes. He drew the central conclusion that a human being, a horse or a tree, ought to be made of the same fundamental substance and that changes in living organisms occur due to the existence of a permanent component.

He refused to accept Plato's subjective interpretation of form. For him form is a concrete event which has its existence in physical individuals.

23.9 Descartes' Soul–Body Dualism Was in Conflict with His Own View of Humans as Machines

After Plato, René Descartes (1596–1650) is considered the most influential of philosophers. The intimate union between mind and body became one of the most puzzling features of his system. He thought that the pineal gland in the brain was where body and mind interacted. The physiological diagrams in his works show how perception and motion of the body are controlled by this gland.

At the same time he held that the mind is distinct from the body and from physical objects. Again, minds could survive the destruction of their corresponding bodies. Minds were considered purely spiritual and radically non-spatial (Lycan 2003). There was a contradiction in Descartes' system because at the same time he considered the human body as a machine.

His location of the mind in a specific part of the brain was a great advancement by emphasizing an experimental approach to the brain.

23.10 Spinoza's Rejection of the Soul and His Emphasis on Nature and Geometry

A few decades later the approach to this problem changed radically. Benedict de Spinoza (1632–1677) was a Dutch rationalist for whom god was identical with nature. Like Descartes he believed that the intellect, rather than the senses, disclosed the essential nature of things and the body and mind were parallel expressions of one reality. He proposed a "geometry of emotions" to indicate that they were ordered processes that in the future would be better understood. He both rejected the idea of a personal god and that of the immortality of the soul (Kenny 2001).

Not until the late 1800s did "soul" and "mind" become separated, soul having an exclusively religious immortal connotation, mind becoming the province of experimental psychology (Bynum et al. 1983).

23.11 Modern Science Started to Investigate the Brain Like Any Other Organ: How to Make Dogs to Salivate at the Sound of a Bell

During the 1800s animal and human physiology had developed greatly but the study of the brain remained a difficult area to approach. The revolutionary spirit that swept Tsarist Russia in the early 1900s had a great impact on science allowing a materialistic approach to the study of the brain. Ivan Pavlov (1849–1936) was a physiologist who devised a radically novel approach to brain function (Fig. 23.3).

The subjects of his experiments were hungry dogs. They were lightly restrained in a sound-proof experimental chamber, and a bell was rung for a number of seconds before the delivery of food. At first the animal would show little reaction to the sound but as the experiment progressed the dog started salivating before the food was presented. The amount of produced saliva was measured by connecting a tube to the salivary glands. Dogs do not normally salivate when they hear a bell, and such a change in behavior was regarded as evidence of learning. The dog learned to expect food. The response was called a conditioned reflex since one stimulus came to stand for another which had originally initiated a reflex response (Pearce 2008).

Fig. 23.3 Diagram of the apparatus used by Pavlov in his study of reflex conditioning with dogs. A hungry dog is located in a sound proof chamber where he can be observed through a *tiny* window. A tube draws the saliva. At 5 min intervals a signal is given: a metronome ticks, a light appears, a musical note is sounded (or a bell rings). The dog learns when to expect food. In this case the saliva starts to form and run. The dog salivated even before the food delivered. This response was defined as a "conditioned reflex" and as evidence of learning

The association of stimuli was considered to result in new temporary neuronal connections in the brain cortex. Pavlov's work had an enormous impact on future research turning animal and human behavior into an experimental science. In the last two decades there has been much experimental work on the processing of visual information, in the brains of cats and monkeys, locating it in the neuronal activity of the cerebral cortex (Eccles 1989).

23.12 Experiments with Humans Indicate "that There Is No Such Thing As a Soul or a Self that's Independent of the Brain"

Following dogs one moved to monkeys and from these to humans.

Yong (2011) describes in detail the experiments carried out by Henrik Ehrsson, a neuroscientist at the Karolinska Institute in Stockholm, Sweden. He uses illusions to probe and displace people's sense of self, using video cameras, goggles and sticks. Two electrodes, on the fingers of the individuals participating in the experiments, record the sweat that automatically erupts on the skin and a computer plots the human's fear on a graph.

The experimental room is stuffed with mannequins, dolls' heads, fake hands, knives and hammers to produce fear. Ehrsson uses these artefacts and virtual reality to create body illusions. In a typical experiment a subject wears goggles showing the view from a camera behind him. An experimenter prods the subject's chest at the same time as prodding at the camera. The result is that the human being sees the hand prodding towards the camera as he feels his chest being prodded. He also sees his body from behind. This creates a vivid sense that his real body is floating behind the one he sees.

Other trials revealed that the self is something that you can dissect. What startled Ehrsson, was the well known rubber-hand illusion. It became possible to convince people that a fake hand was their own by hiding their real hand under a table, placing a rubber one in front of them, and stroking both in the same way. The result was a surreal sensation.

Illusions, like this one, work for around four out of five people, volunteers show a reflexive burst of nervous sweating. By improving his experiments he even convinced people that they owned a third hand. His illusions led the human brain to think that it was somewhere else entirely.

Such illusions are thought to be dependent on "multisensory" neurons that combine vision with touch. Their location in the brain has been attempted by inducing illusions while volunteers sit inside a functional magnetic resonance imaging machine. They are located in the cortex but their exact site remains to be determined.

One of his colleagues, at Mainz University, Germany, Thomas Metzinger did not hesitate to state: Ehrsson's work speaks to the idea "that there is no such thing as a soul or a self that's independent of the brain".

Sources of Figures

Fig. 23.1 Cuvillier, A. 1941. Manuel de Philosophie, Vol. 2. Librairie Armand Colin, Paris, France (Fig. 80, page 606).

Fig. 23.2 Cuvillier, A. 1941. Manuel de Philosophie, Vol. 2. Librairie Armand Colin, Paris, France (Fig. 81, page 607).

Fig. 23.3 Pearce, J.M. 2008. Animal Learning and Cognition. Psychology Press, Taylor and Francis Group, New York, U.S.A. (Fig. 1.8, page 29).

Brain Imagery: Visual Perception of Form, Color and Motion 24

24.1 Three Parallel Pathways Process Visual Information Involved in Form, Color and Motion

The anatomical components of the visual pathways are: the optic nerve (1), that projects to the thalamus (2), which in turn projects to the visual cortex (3) (also called striate cortex). As a result of this orderly connection the striate cortex contains a complete map of the retina. Beyond the striate cortex lies the extrastriate cortex also containing information from the retina. This is divided into 32 representation areas which differ in their functions: *V5* is primarily concerned with *motion*, while *V4* is much more concerned with *color*. *Form* is analysed by the parvocellular-interblob system (Kandel et al. 1995).

24.2 The Perception of Form by the Brain: Imagery Transforms the Received Image—The Outer World Is Seen Upside Down

On inspection with the naked eye the brain appears as a rather irregular structure. Its outer surface is covered with circumvolutions that make it look like a labyrinth. This has led to think that its millions of cells, have been put together randomly, have shapes not necessarily related to their functions and are aggregated into areas that do not communicate with one another. But appearances delude.

The experimental studies carried out in recent years have superseded this view. Actually, the vertebrate brain integrates visual information through several parallel and well-defined pathways. Its architecture plays a critical role in the segregation and processing of visual information (Hill et al. 2004).

From the very beginning the brain is obliged to transform reality.

To start with, the retinal image is an inversion of the visual field. The brain gets a picture of the outer world which is upside down but it contains a mechanism that adjusts this inversion.

Additionally the optic nerves from each eye join at what is called the *optic chiasma*. At this location the fibers from the part of the retina, close to the nose, cross to the opposite side, whereas the axons of ganglion cells, in the part of the retina far from the nose, do not cross. This separation forms the left and right optic tracts. These tracts project to three subcortical targets but only the lateral geniculate nucleus processes visual information connected with the perception of form. The result is that this area of the brain contains a *neural map* of the retina and is the site where the processing of vision information takes place.

The brain has an impressive ability to sort out different inputs. The primary visual cortex in each cerebral hemisphere receives information exclusively from the contralateral half of the visual field and it consists of six cell layers with distinct appearances. Besides, this area of the cortex contains two basic classes of cells: pyramidal and nonpyramidal, which are large and small respectively and have different functions. However, a central problem demands elucidation: How are the receptive fields of visual neurons transformed in the cerebral cortex? Unexpectedly, it is the cell's shape and internal organization that is a decisive factor.

24.3 Brain Cell Geometry at the Basis of Form Perception

Among the pyramidal cells there are *simple* and *complex* ones. They not only have a specific orientation but the linear stimulus must also occupy a specific position within their receptive field to be optimally effective. Besides, these fields may be circular or rectilinear. Simple and complex cells respond best to linear stimuli with particular orientations. These two cell types are considered to analyse form because they are sensitive to the outline of an object, to its contours, boundaries and contrasts (Hubel and Wiesel 1962; Robertson 2009).

A. Lima-de-Faria, *Molecular Origins of Brain and Body Geometry*,
DOI 10.1007/978-3-319-06056-9_24, © Springer International Publishing Switzerland 2014

This information is reinforced by the finding that the initial steps of perception are accompanied by a series of cell interactions in which one group surveys the action of another which in turn surveys the next one. Complex cells survey simple ones, which survey stellate cells, which control the activity of geniculate cells, which again survey retinal ganglion cells. At each level every cell has a greater capacity for abstraction than the cells of the lower level.

Besides, some cells respond only to horizontally oriented stimuli. Neurons within a column respond to a single axis of orientation and column cells may have different axes of orientation. Additionally, the cell columns in the primary visual cortex communicate with one another by means of horizontal connections (Kandel et al. 1995).

We are confronted with geometry being part of the physical basis of form perception.

The metabolic activity of the different types of cells was monitored by using radioisotope labeled molecules, such as glucose, injected into the blood stream of monkeys.

Hence, molecular mechanisms, cell geometries and cell communication are opening the way to an understanding of the perception of form. But we are far from understanding how the upside down view of the world received by the retina is transformed into the correct reverse position present in reality. This is a key feature of brain imagery that cannot be denied.

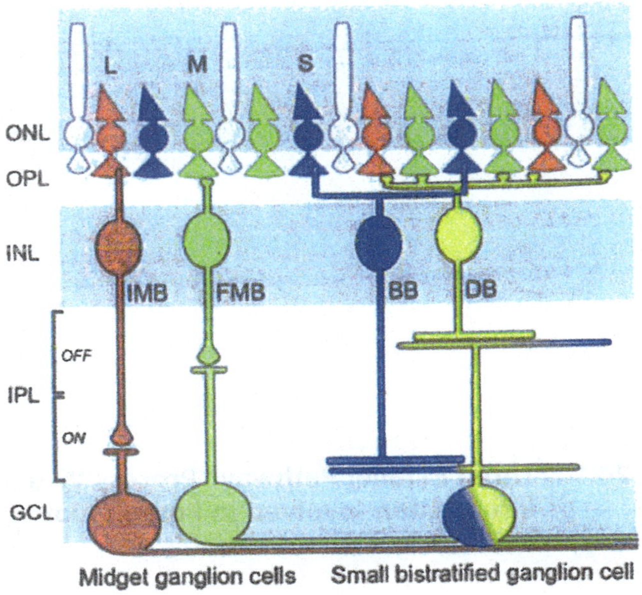

Fig. 24.1 Color-coded pathways in the primate retina. In the central primate retina, invaginating ON midget bipolar cells (IMB) and FMB cells are each connected to one long (L) or middle (M) wavelength-sensitive cone only and they innervate individual midget ganglion cells. Hence, they form a *color-coded private-line* transmission of cone signals to the brain. The small bistratified ganglion cell receives ON input from short (S) wavelength-sensitive cones via *blue cone* bipolar cells (BB) and OFF signals from diffuse bipolar cells (DB). The latter are mainly connected to M- and L-cones. Therefore, they provide *yellow*-OFF input for the small bistratified cell. ONL/INL, outer/inner nuclear layer; GCL, ganglion cell layer

24.4 The Perception of Color: A Single Photon Is Enough to Initiate Visual Perception

Photons are the constituents of electromagnetic radiation which may take the form of light, X-rays and gamma rays. They are elementary particles having a discrete energy but no mass (Lincoln 2012). Accordingly their energy, like that of other elementary particles is minimal. Yet the cells of the human eye are so sensitive, due to their molecular organization, that a single photon produces (even without mass) a visual effect.

It was Hetcht (1942) who showed, by measuring the input of light during day and night vision, that rods (the cells of the retina) can be excited by a single photon.

24.5 Vision Involves Molecular Cascades Which Are Initiated and Terminated by Specific Proteins

Visual perception begins by photons being projected onto the back of the eye, where they are converted into electrical signals by the retina. The signals are sent by the optic nerve to the brain, leading to perception.

The retina is not a peripheral organ but a part of the central nervous system like the brain. Its cells are neurons which are specialized in converting light into electricity because photon absorption leads to electron emission.

The human retina contains two types of photoreceptors: rods and cones. They are responsible for night and day vision respectively. Cones also mediate color vision but rods do not participate in this event. There are three types of cones each with a distinct pigment that is most sensitive to a different part of the visible spectrum. Rods and cones are filled with pigments each consisting of a small light-absorbing molecule attached to a large membrane protein. The protein portion does not absorb light. It is the pigment part, *retinal* (a derivative of vitamin A), that by assuming different molecular configurations (within minutes), is responsible for phototransduction (Fig. 24.1).

Photoactivation is characterized by super speed and amplification. A single *rhodopsin* molecule (the protein which contains retinal) can activate more than 10^5 molecules of a second messenger molecule in 1 s. This extremely rapid amplification is achieved with the intervention of a regulatory protein called *transducin* (Kandel et al. 1995).

Molecular order decides the end of visual perception. The molecular cascade initiated by *rhodopsin* is terminated by

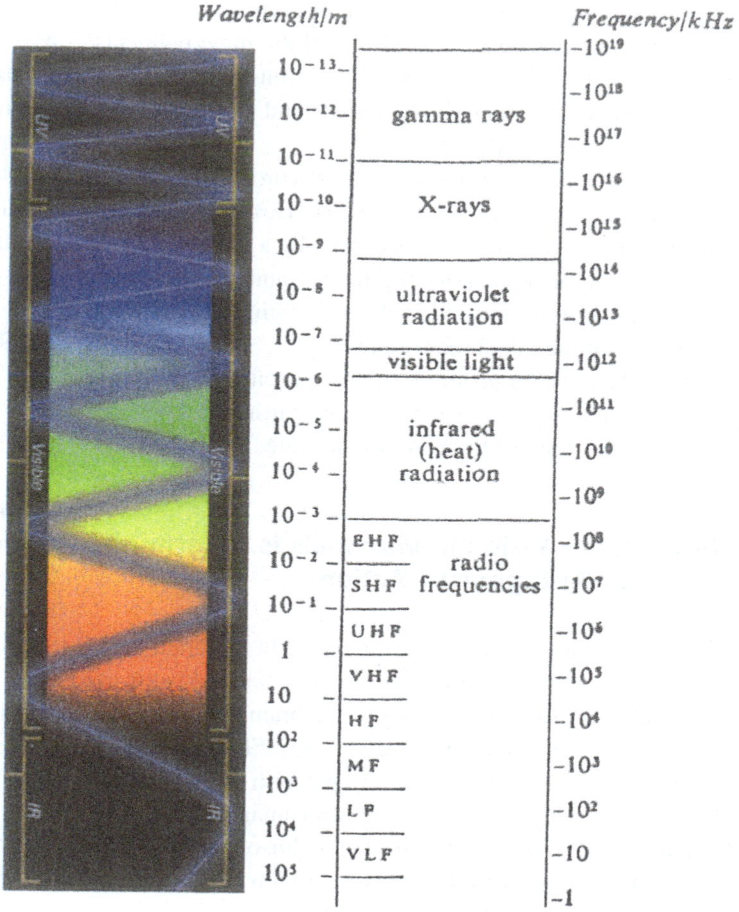

Fig. 24.2 **Light waves of different wavelengths, each equating to different colors to humans** (short wavelengths appear as *blues*; long wavelengths appear as *reds*). **Spectrum of electromagnetic radiation.** Waves of energy are caused by the acceleration of charged particles. They consist of electric and magnetic fields. The characteristics of electromagnetic radiation depend on the frequency of the wave motion which is related to the wavelength by a simple equation. The lowest frequencies are associated with radio waves, the highest with gamma rays

another specific regulatory protein called *arrestin* leading to its rapid inactivation. The output of the retina is conveyed to ganglion cells which in turn project to the brain. During this process the electrical signals are combined depending on the precise spatial and temporal patterns of the light that stimulates the retina.

24.6 Photons of Different Wavelengths Exist, but Color Is Another Invention of the Brain

Visible light represents a tiny fraction of the enormous range of wavelengths that characterize electromagnetic radiation. The wavelength varies from 10^{-14} m in cosmic rays to several meters (radio waves). White light is the name given to wavelengths of the visible spectrum. It extends from 700 to 400 nm. On both sides are the infrared and the ultraviolet regions. Photons are the carriers of all wavelengths (Fig. 24.2). Thus, what exists in nature are photons with different energies not colors. Significant is that the spectrum has a continuous variation in wavelength but seven colors are usually distinguished: violet, indigo, blue, green yellow, orange and red and within sharp limits (Pitt 1988).

Like the perception of form, color perception involves a sophisticated abstracting process—actually another type of imagery.

The recognition of simple brightness is different from color perception. We can discriminate between 500 gradations of brightness but we can distinguish more than seven million gradations of color. This is why books, used as references to describe the colors of flowers or other organs of living organisms, do not use names, but only numbers, to identify the different color nuances.

As we all have experienced paint or clothing have different colors when inspected in artificial light or in daylight. A white object can appear pink or pale green against different backgrounds.

The immediate question is how reliable is the brain?

Colors do not all have the same value. There are three primary colors: blue, green and red. This selectivity of the brain called *trivariancy* is actually the basis of color television, which uses three types of phosphors to reconstruct the full spectrum of perceived colors. The whole visible color spectrum is synthesized from only three basic colors by the brain imagery.

The *trivariancy* has a physico-chemical basis, it results from three types of light-absorbing cone photoreceptors each with a different visual pigment present in the human retina. A single cone cell contains only one of the three pigments. One pigment is primarily sensitive to blue (short wavelength). Another is responsive to green and a third is receptive to red (long wavelength).

Color blindness can be the result of cortical lesions, but also be due to disease of the retina. However the most common form (red or green) is caused by a mutation of the human X chromosome affecting the red and green pigment genes. Mutations also occur in the blue pigment gene which is situated on the 7th chromosome.

How the cortex transforms a photon excitation into color remains a question mark. We are only starting to unveil the enormous capacity of the brain to innovate.

24.7 Color Vision in Mammals Is Determined by Antagonistic Action

Three different pathways are established in the brain based on the wavelength of the radiation. These parallel pathways reach different types of neurons in the brain. The combination of excitatory *blue-ON* and *yellow-OFF* input creates color opponent signals which take place in ganglion cells. This color opponency is achieved by an antagonistic organization of the brain. Proof of the color-opponent signal processing was obtained by applying immunohistochemical methods and confocal microscopy (Puller and Haverkamp 2011).

24.8 The Evolution of Pigment Genes

Animals are dichromatic, trichromatic and tetrachromatic if they can recognize two, three or four separate basic colors respectively.

Opsin is the general term given to proteins of the rhodopsin family. Vertebrates have five classes of opsins: A rod (*Rh1*) class restricted to rod photoreceptors and four different cone classes distinguished on the basis of their sensitivity to the wavelength of radiation and their amino acid sequence. The four classes of cone opsins in photoreceptors are divided according to their wavelength sensitivity: *LWS* longwave-sensitive (500–570 nm), *MWS* middlewave-sensitive (480–530 nm), *SWS2* shortwave-sensitive (400–470 nm) and *SWS1* also shortwave-sensitive (355–445 nm) (Hunt et al. 2009).

The evolution of visual pigments has been determined by studying gene sequences. Cone pigments preceded rod pigments. Mammals, which arose around 150–200 million years ago, have only two classes *Rh1* and *LWS*. In marsupial and eutherian mammals there is a loss of the *SWS2* and *Rh2* genes. The monotremes (*platypus*) that diverged from the marsupial/placental lineage around 200 million years ago have discarded the *SWS1* gene but retained the *SWS2* gene. Primates have an improved color discrimination in the red/green region of the spectrum, that led them to become trichromatic.

Most aquatic mammals have dispensed with color vision through the loss of a functional *SWS1* gene while retaining an intact *LWS* gene.

24.9 More than a Thousand Opsins Have Been Identified. At the Same Time There Has Been a Loss of Opsin Genes During Evolution

Terakita (2005) has reviewed the extensive information on opsin molecules. Like many other proteins they have been crystallized. This has allowed the resolution of their atomic structure and configuration. They were also amino acid sequenced. As a consequence more than a thousand opsin genes have been identified.

Their ancestry goes back to the fungi, green algae and bacteria, since in these simpler organisms there are opsins which function as light sensors. These have different amino acid sequences but the same three-dimensional structure (Terakita 2005).

Transgenic mice, as well as different mutants, have been used to identify and localize the numerous genes. There has been a loss of opsin genes during animal evolution. The fruit fly *Drosophila* has seven and humans have 9 opsin genes. The molecular organization as well as the exact chromosome location of these genes has been established.

24.10 Music Tones Are Translated into Color and There Is Also a Relationship Between Letters and Color

Alexander Scriabin (1872–1915) was a Russian composer, who, like some other persons, had an extremely intense sensation in which different musical tones were immediately translated into color. He experienced all music as a rich combination of inner colors. His composition "Prometheus, the Poem of Fire" contained a part for "color organ" (Salter 1978).

Scriabin was a student of Nicolaj Rimskij-Korsakow (1844–1908). When they listened to the same concert, a D-tone was yellow for both Scriabin and Rimskij-Korsakow. Listening to other music they had different opinions: a tone being green to one and violet to the other. Scriabin went as far as to play piano using lamps that lighted the concert hall with different colors which fitted the different parts of his music (Vester 2004).

There is also a relationship between color and letters. The American physicist and Nobel laureate Richard Feynman (1918–1988) described in his books that when he wrote equations he saw the letters in color. Other individuals who were tested by psychologists could see letters (and even the days of the week and of the month) as represented by colors. The French poet Arthur Rimbaud (1854–1891), in one of his poems, expressed it clearly: *A* is black, *E* white, *Y* green and *O* blue (Vester 2004).

The relationship music—color was partly elucidated by studying the behavior of an American patient. He was a painter who became color blind in an accident. Suddenly he saw the world around him as a gray environment. Both color and music were impaired for him. Special scan techniques were used to study his brain damage and especially the visual cortex. It was clear that he could discriminate wavelengths but he could not translate them into color perception showing that the two events were separated (Kandel et al. 1995).

The brain not only created color as a separate process from wavelength excitation but also translated music and letters into color displaying additional forms of imagery.

24.11 The Evolution of Color Vision Was Disclosed by Studying Monkeys and Their Genes

Every physiological process has evolved, since the genes responsible for its functions have changed with time. Like any other atomic construction DNA evolves.

Neumeyer (2012) starts describing the evolution of color vision in vertebrates by stating that it is an old invention, since the ability to distinguish the three basic colors exists in fishes, amphibia, reptiles and birds. Suddenly this form of vision became widely absent as the mammals emerged but "was reinvented by Old World primates". This conclusion was confirmed by an analysis of opsin genes in monkeys. The event occurred about 30 million years ago when New World monkeys, which have only two pigments, separated from Old World monkeys which have all three pigments.

The brain had also to adapt to the variation in gene function which it did not control.

24.12 There Is Real Motion and Apparent Motion: The Brain Invents an Abstract Form of Motion: Yet This Can Be Located in a Specific Brain Area

The visual system has two ways of detecting motion: one based on the real motion of the image and the other on the movement of the head and the eyes. Only humans and primates can respond to objects that do not move. On the other hand frogs and deer cannot see objects unless they move (Yarbus 1967; Kandel et al. 1995).

Well-known phenomena expose the imagery of the brain when it deals with motion.

When we watch movies, most of the time our eyes experience a constant movement. If at the end of the show we inspect the film from which the pictures were projected on the screen no movement is to be found on the film. It only consists of separate still images. This illusory movement was invented by the brain and is called *apparent motion*.

Real motion in the visual field can be detected by comparing the position of images perceived at different times being able to resolve events separated in time by a few tenths of milliseconds.

Apparent motion is more convincing than actual motion. In a simple experiment, that is of the same type as film projection, lights are separated in space and turned on and off at appropriate intervals giving the illusion of motion even though the lights do not change position. This constitutes evidence of the existence of a special visual system involved in the detection of motion.

Experiments with monkeys led to the finding that motion is represented in two temporal areas of the brain. The brain seems to compute the loci of possible motion in which the cortex neurons are integrated by higher-order neurons.

Sources of Figures

Fig. 24.1 Puller, C. and Haverkamp, S. 2011. Bipolar cell pathways for color vision in non-primate dichromats. Visual Neuroscience 28: 51–60 (Fig. 1, page 52).

Fig. 24.2 Parker, A. 2004. In the Blink of an Eye. Free Press, London, UK (Fig. 11 on Plate). Pitt, V.H. (Editor) 1988. The Penguin Dictionary of Physics. Penguin Books, London, UK (Table 7, page 424).

25 The Dream: Another Form of Brain Imagery

25.1 "The Major Goal of Cognitive Neural Science Is to Study Internal Representations of Mental Events"

The above statement is the starting sentence, written by Kandel et al. (1995), in their chapter: "From Nerve Cells to Cognition".

According to them cognitive neural science relies on several main sources of information.

1. The brain has an orderly representation of personal space. This is due to the existence of reliable cortical maps of the human body.
2. The cortex has also a map of the body for each sensation but different species of animals rely on different parts of the body for sensory information.
3. The brain perceives the body in its totality. If the whole pattern of the human body is altered by amputation of one of its members, the brain persists in identifying it as if it were still present. This is revealed by the "Phantom Limb Syndrome" in which a sensation of the hand and pain persist following amputation. Thus, the internal representation of personal space is modifiable by experiment.
4. Mental events can be studied at the cellular level since each central neuron has a specific receptive field. This is demonstrated by inputs to the somatic sensory cortex which is organized into columns that are integrated by cells having feature-detecting properties.
5. Real (as well as imagined and remembered) space is represented in specific brain areas. Lesions of the cortex lead to defective behavior which is dependent on the activity of single nerve cells located in the posterior parietal cortex.
6. The human brain consists of a 100 billion nerve cells. But by studying one brain cell at a time, it became evident that information received in various regions of the brain is transformed into perception and planned action. Each central neuron has a specific field of receptivity, yet cells integrate information.

The picture that emerges is of an unexpected but impressive order. The brain has its own "reliable cortical maps".

This information is the result of extensive research on the brains of intact and active primates (including humans) by using an array of novel techniques: positron emission tomography, magnetic resonance imaging, magnetoencephalography and voltage-sensitive dyes.

25.2 Visual Pathways Disclose that the Brain Does Not Simply Receive Pictures but Rather Constructs Visual Images

Visual perception has turned out to be a creative process. Initially, vision was considered to be like the operation of a camera. The lens of the eye, like the lens of a camera, focuses an inverted image onto the retina. This analogy was too simplistic because vision really creates a three-dimensional perception of the world, that is different from the two-dimensional images projected onto the retina. In addition, we are able to recognize an object as the same even though the actual image on the retina varies widely under different light conditions. When people walk toward you, you perceive them as coming closer, you do not perceive them as growing larger, even though the image on the retina enlarges. The visual system does not record images passively, like a camera. Instead it transforms light on the retina into a mental construction by a creative process (Kandel et al. 1995).

The process of perception actively shapes, from the details of a stimulus, the complete form that emerges in consciousness. The perceived image is not the sum of its elements, rather the different components are selectively organized by our brain creating a form that is more than the sum of its parts.

Psychologists have been led to consider laws of perception. There is a tendency to perceive one pattern rather than another. These laws are innate and govern the pattern, shape, color, distance and movement of objects in the visual field. Perceptual organization is continuous and dynamic. A well-known example is the alternation of figures on a background.

A. Lima-de-Faria, *Molecular Origins of Brain and Body Geometry*,
DOI 10.1007/978-3-319-06056-9_25, © Springer International Publishing Switzerland 2014

The image can be seen as two black profiles or as a white vase, but it is almost impossible to see both images simultaneously. Many other figures have since then been added which demonstrate the creative filling by the brain in the process of vision.

25.3 Sleep Evolved Like Any Other Physiological Process: From Fish to Mammals Primary Sleep Became Transformed into True Sleep

Originally sleep was treated as a particular case of behavior being a source of speculation. In the last four decades it became an area of active research involving different experimental methods: creation of lesions in the brain, implantation of electrodes in the forebrain and optic region, study of eye movements, respiration, heart rate and motor activity.

Simultaneously, the study of sleep that had been confined mainly to birds and mammals, was extended to all vertebrates starting with fishes.

After the discovery of rapid eye movement (REM) phase of sleep in higher vertebrates (birds and mammals), the "state of rest" was analysed in fish by implanting electrodes in the neck muscles (electromyogram *EMG*) and in the forebrain (electrogram *EG*). Changes in the bioelectrical activity in the forebrain did not occur during the transition from wakefulness to rest. But rapid eye movements (which are characteristic of a phase of sleep) were detected in freshwater fish occurring during the rest phase.

Subsequently the alternation of wakefulness and rest were studied in amphibians (tree frogs). Following the same procedure electrodes were implanted into the midbrain and some muscles. Night time states of rest were not accompanied by an increase in bioelectrical activity. A comparable experiment with salamanders gave the same results.

The sleep of reptiles has been investigated extensively in the last years in: chameleons, iguanas and turtles.

The advent of the reptilian brain led to a big jump in sleep evolution. Recordings of the bioelectric activity in the forebrain and the neck muscles, as well as heart rate, led to distinguish three states: wakefulness, resting and sleep. Characteristic features of sleep in these species were: eye movements, deceleration of heart rate, reduction of muscle tone and lower bioelectrical activity amplitude during wakefulness. All authors agree that sleep is present in reptiles.

In birds and mammals sleep has been divided into two main stages: (1) Paradoxical (*PS*) or Rapid Eye Movement (*REM*) and (2) Slow Wave Sleep (*SWS*) also called Non-Rapid Eye Movement *NREM*. In hens and pigeons the SWS phase is characterized by: increase in the amplitude of slow bioelectrical activity and drop in heart rate, whereas *PS* shows a still more pronounced muscular tone decrease and a series of rapid eye movements. *PS* in birds amounts to about 0.3 % of the overall duration of sleep (Karmanova 1982).

The combined information from the paleontological and physiological data led to the conclusion that *SWS* sleep in mammals appeared about 180 million years ago, whereas *PS* appeared 50 million years later. *PS* is already evident in marsupials (opossum) (Hobson 2009) (Fig. 25.1). As a consequence the evolution of sleep was divided into three main periods: primary, intermediate and true sleep. Fishes and amphibia show rest forms characteristic of primary sleep. They have a night-time form of rest (SLS) which is considered the antecedent of intermediate sleep in reptiles that later in evolution led to *SWS* and *PS* phases in birds and mammals (Fig. 25.2).

25.4 The Dream Is a Form of Imagery that Occurs During Sleep

The information on sleep leads directly to the study of dreaming which occurs during this physiological activity. The two events are directly related because dreams occur every night in every sleeping person. During this process the brain is disconnected from the environment and generates an entire world of conscious experiences by itself.

The combination of brain lesions, functional imaging and neurophysiology have opened a new era in which the dream emerges as closely related to mental imagery and to perception (Nir and Tononi 2009).

Sigmund Freud (1856–1939), one of the initiators of the study of dream, predicted that future research would discover an organic basis for this mental event, a prediction that is now being fulfilled.

From the beginning dream content was much used in psychoanalysis but modern dream research tries to relate the activity of neurons to dream form rather than content.

The level and nature of conscious experience varies dramatically during sleep. At present sleep is divided into several phases which are related to dream in different ways. (1) The early phase, called SWS for slow wave sleep occurs as night begins and is characterized by many large slow waves in the electroencephalogram. During this period consciousness can nearly vanish, despite continued neural activity. (2) An intermediate stage between awake and deep sleep with sharp waves and slow eye movements. (3) Sleep occurring throughout the night characterized by oscillations originating in the thalamus. (4) Late at night sleep characterized by muscle tone and involuntary rapid eye movements (*REM*).

Subjects awakened from all phases of sleep report full-fledged dreams consisting of vivid hallucinatory experiences that tend to follow a narrative. "The brain is creating a story, filling it with actors and scenarios and generating

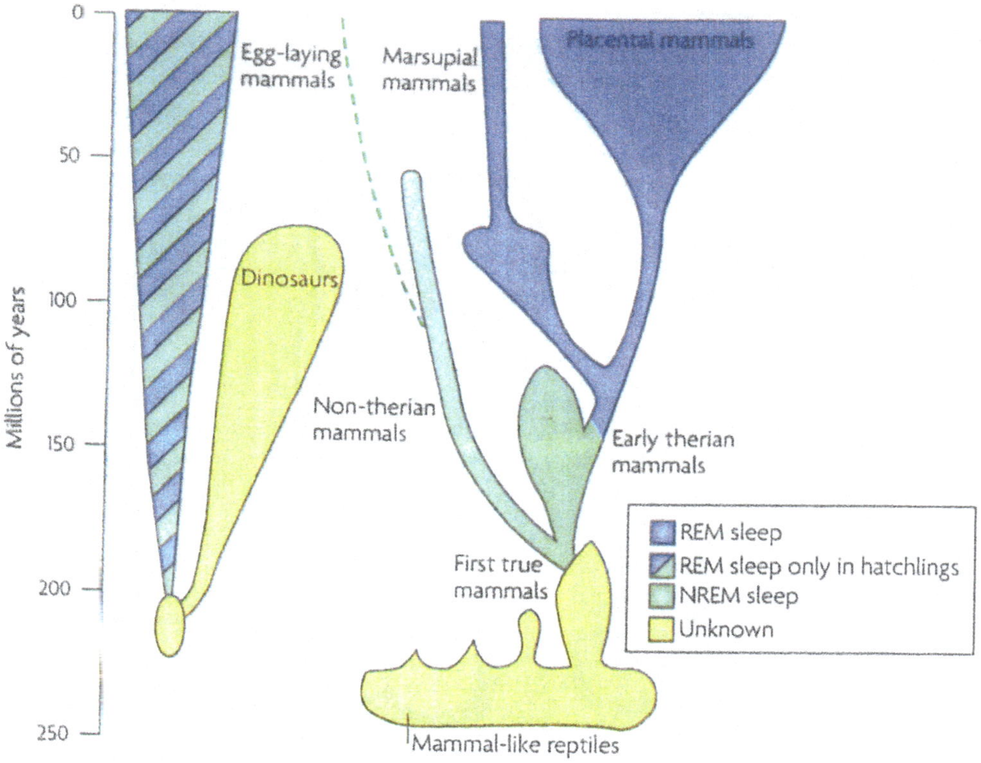

Fig. 25.1 **Evolution of rapid eye movement sleep (*REM*).** Birds and mammals evolved separately after branching off from the ancestral tree many millions of years ago. Both birds and mammals are homeothermic, and both classes of animal have appreciable cognitive competence. Restoration of thermoregulatory capacity is a strongly evidenced function of REM sleep. With respect to the possible enhancement of cognitive skills by REM, it is significant that both birds and mammals are capable of problem solving and both can generate forms of verbal communication

hallucinatory images" (Nir and Tononi 2009). The experiments emphasize that dreams are not created from a vacuum but closely reflect the organization and functions of the different parts of the brain.

Perception in dreams follows closely wakefulness. Dreams are highly visual many times including color, sound and movement as well as taste, pleasure and pain.

Besides, the dream is not a radically different situation from wakefulness. Not only the electroencephalograms look similar in active waking and rapid eye movement sleep but positron emission tomography also revealed that global brain metabolism is comparable between wakefulness and this latter type of sleep. In agreement is the finding that patients with brain lesions that impair their waking cognition show corresponding deficits in dreams.

Dreams also differ from waking experience in various ways. (1) During dreaming there is a reduction of voluntary control of action and thought. One has no control over the content of dreams. The right parietal cortex was found to be deactivated during the most active phase of dreaming. (2) A dreamer easily accepts impossible events such as flying or seeing a pink elephant. A process that seems to be related to the deactivation of various brain regions. (3) Surprise, anger, fear and anxiety are typical emotional stages of many dreams. These have been correlated with marked activation of limbic and other brain structures. (4) The dream alters memory, most dreams are lost forever unless written down. Moreover, a dream is not like an episode of life being replayed, but "residues" from experienced events are incorporated in most dreams.

There are a series of contemporary hypotheses on what dreams represent such as a fulfillment of unconscious wishes which are actively repressed. These are mainly of interest to psychoanalysis and do not concern us here. Instead as neurobiology progresses more information becomes available on the chemistry of dreams. Dreaming is driven by the dopaminergic system as revealed by patients operated on the brain (prefrontal leucotomy) and the effects of *L-DOPA* (which is the precursor of the neurotransmitters dopamine and norepinephrine (Solms 2000). Administration of pilocarpine can induce an artificial rapid eye movement sleep period associated with dreams (Gillin et al. 1985). But dreaming is considered to be driven by a series of specific chemicals and brain regions where serotonin, norepinephrine and histamine are reduced, or absent, whereas acetylcholine and dopamine are both present.

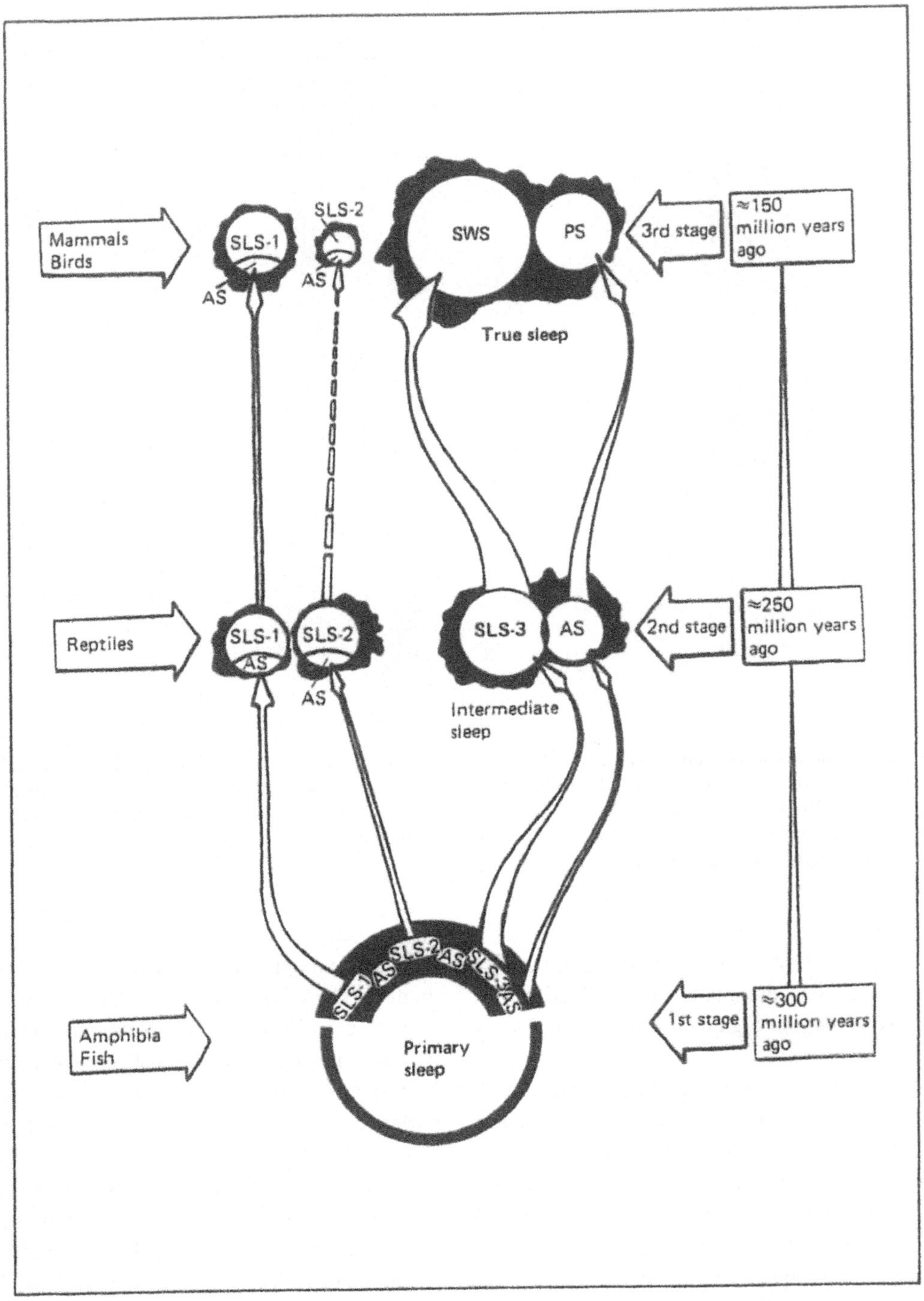

Fig. 25.2 **Scheme of transformation of primary sleep rest forms (SLS-1, SLS-2, SLS-3) in vertebrates.** AS = Activation, spontaneously arising against the background of all forms of rest of primary and intermediate sleep

25.5 In Dreams One Sees Colors and These Are the Same that Are Invented During Wakefulness: Color Without Photons

During dreaming we have clear images that are painted in the most vivid colors. These are experienced although our eyes do not receive photons from the outer environment. Thus, without photons reaching the retina, since our eyes are closed and rooms are dark during sleep, the same imagery of colour occurs in the brain.

This may be due to storage of the action of light on cells, just as other components of the dream are a recapitulation of earlier experienced situations. Besides, the colors seen during sleep are not novel ones, but the same that are experienced during wakefulness, an indication that the same brain channels and pathways are being used.

25.6 Artists Often Depict Situations Long Before Scientists Arrive at the Same Conclusion: The Painting of Dreams

Great artists, poets and musicians, in their deep search for the central problems of life, may arrive at the concretization of a thought that expresses a novel situation in an accurate way.

Scientists are many times baffled by finding that non-scientists reached their conclusions, at an earlier period, without having used their elaborate tools and painstaking work. Scientists, like artists used intelligence and intuition in their endeavours, but observation and experiment, are their main sources of guidance. As such it takes much longer in science to come to a similar final result.

The examples of this situation are many. J.A. Hobson, from the Harvard Medical School, could not refrain from printing, in his article in *Nature* (2009), dealing with sleep and dreaming, the painting by Salvador Dali entitled "Dream caused by the flight of a bee around a pomegranate a second before awakening" (painted in 1944).

Dali is acknowledged as the leading painter of the artistic movement called Surrealism, because where others expressed dreams in a vague form, without particular reference to their construction or origin, the Spanish artist painted the dream in an unprecedented and masterly way.

Hobson expresses his amazement of how Dali's work "anticipates modern dream science findings in several ways". According to him, these are: (1) The representation of vivid and visual imagery. (2) The dreamer is deluded into believing that he is awake. (3) The loosely association of bizarre and discordant images. (4) These are bound together in a coherent but fleetingly remembered narrative. He concludes that with the help of modern neuroscience many of the painting features can begin to be "explained naturalistically without recourse to questionable psychological speculations". It took 70 years.

Sources of Figures

Fig. 25.1 Hobson, J.A. 2009. *REM* sleep and dreaming: towards a theory of protoconsciousness. Nature Reviews/Neuroscience 10: 803–813 (Fig. 2, page 806).

Fig. 25.2 Karmanova, I.G. 1982. Evolution of Sleep. Karger, Basel, Switzerland (Fig. 22, page 63). Nauka Publishers, Moscow, Soviet Union.

Plato Had No Electron Microscope: Molecular Geometry Reverses His Concept of Reality

26

26.1 Antiquated Terms Had to Be Substituted to Allow New Ideas

Science is based not only on the shedding of old concepts but also on the discarding of terms that were connected with them. This is why references to words denoting: instinct, intelligence, determinism and others, which dominated during the early 1900s were substituted or abandoned in the last decades. Instead cognition, consciousness and behavior are used at present (Lazareva et al. 2012).

Cognition is defined as referring "to the mechanisms by which animals acquire, process, store, and act on information from the environment". It includes perception, learning, memory and decision-making.

There are two major differences from the earlier approach. One, is based on the finding that animals cannot be separated from humans in their mental processes. The other, is that these processes have a common denominator—they are reducible to cellular and molecular events in every brain. Cognition is mainly an event in which animals including humans process information acquired by the senses (Shettleworth 1998).

26.2 A Cloned Einstein Would Not Necessarily Conceive Relativity: Several Major Factors Participate in the Emergence of a Scientific Concept

To realize the gap between the way of thinking in Plato's time and at present, it is necessary to understand how concepts arise in science.

Several major factors participate in the birth of a scientific idea, or any other creative process, such as music or painting:

1. The personality of every individual. This is a product of the genetic constitution and of the environment faced during pregnancy, infancy and adolescence.
2. The historic moment. A major component is the quality of learned schools where knowledge is acquired, as well as the type of colleagues active in a given scientific field. Another critical component is the dominating political ideology.
3. The economic development. It may be a situation of poverty or it may allow the accumulation of wealth permitting a non-profitable behavior based on a surplus.
4. The technological advancement. This empowers humans with tools allowing them to reach novel levels of the real world. There were no eye glasses, microscopes or telescopes in the ancient world. Glasses started to appear in the 1500s and microscopes and telescopes were invented in the 1600s. They allowed Leeuwenhoek and Galileo to reach a hitherto forbidden level of reality.

It is not science fiction, but present day biotechnology, could, in principle, produce a new Einstein by cloning. One could use the same technique that is employed at present to recover a living mammoth or cloning mice (Greewood et al. 1999; Wakayama et al. 2008). Starting from a few cells of his mouth, left in his tooth brush, an Einstein born in our days would not have the same loving parents, an attentive wife, but above all, he would lack the unique colleagues with whom he collaborated in finding out the equations that define relativity. Among his eminent colleagues were the Indian physicist Satyendra Bose, Max Planck, Niels Bohr and Max Born. A cloned Einstein would be immersed in a society lacking the intellectual ferment of the early 1900s that demanded a revision of every accepted idea both in society and in science. He would be utterly confused and disgusted.

The same would be the case with Plato. The pleasant, quiet and abstract world, of everyday Athens, could not be reconciled with the reality of today's interplanetary travel. Besides, Socrates, an indispensable colleague, was not living next door. If cloned, Plato would also retract in disgust and amazement. This was not the world, based on simple

A. Lima-de-Faria, *Molecular Origins of Brain and Body Geometry*,
DOI 10.1007/978-3-319-06056-9_26, © Springer International Publishing Switzerland 2014

technology and accepted slavery, on which his ideas had been cemented.

26.3 Plato's Great Contribution: The Separation Between Unreal and Real Worlds. For Him Ideal Forms Were Unchanging and More Real than Objects

It is well known that "the first casualty of war is truth". Every contender usually claims that he is winning the battle. The same applies to those who became well-known for a novel idea. They pay the price of being, not only misquoted, but even accused of statements that they never uttered. This is obviously the case with Plato who has been, time and again, cited for over 2,000 years.

His work covered several areas concerning the reality of the outer world.

To start with, as Hornblower and Spawforth (1999) emphasize "Plato tries to stimulate thought rather than to hand over doctrines". But his ideas are uncompromising in several respects.

A person's soul is an entity distinct from the living embodied person. The soul is the principle of life and of the intellect. Souls are immortal and are endlessly reborn into different bodies.

When it comes to the understanding of reality, the senses give us merely unreflected and unreliable reports. He outlined what has since been called the theory of "forms", which is linked to his concept of the immortality of the soul. It is a pivotal concept.

For Plato, forms exist independently of the material world. They are independent, real, invisible and changeless (Blackburn 1996; Brunnin and Tsui-James 2003). Forms "are unchanging objects existing apart from this world and more real than it, knowable by the mind as opposed to the senses" (O'Connor 1985).

In geometry he distinguished between the notion of perfect triangles and perfect circles as opposed to the diagrams which we draw on paper. Drawn triangles are only approximately triangular, all have minor flaws. In contrast, the triangles that are the objects of our geometry theorems are necessarily perfect. They only exist in our minds. "Even the most beautiful woman or statue is not flawless". For Plato absolute beauty cannot exist in the world of sense (O'Connor 1985). Besides, the empiric outside world has no independent existence, it only exists to the extent that the objects agree with or are part of immortal ideas (Delius et al. 2000).

26.4 In Plato's Time Democritus Conceived the Existence of Atoms but the Idea Was Rejected Because It Did Not Fit His Concept of an Abstract World

The partition of living organisms and of matter into minor components was a thought well-established in Plato's time.

Democritus (born about 460 B.C.) is mainly credited with the atomic theory of the universe which he expounded in great detail. Atoms were indivisible units, that were solid, and invisibly small. They differed from each other in shape and size and in their relative and absolute position through movement in space. By their changes of position the atoms produced the compounds of the changing sensible world. These compounds differed in quality according to the shape and arrangement of the component atoms. Some of these atoms were sorted out producing distinct masses having the appearance of earth, water, air and fire (Hornblower and Spawforth 1999).

The concept was so well-defined and logically conceived that the description agrees with what at present is called a chemical reaction.

The work of Democritus covered 70 titles but none survives. This may not be an accident. As Hornblower and Spawforth report: "Plato although he must have known his work, never mentioned him by name". Silence is the best weapon used in science and the destruction of books has been for ages the most effective way of disposing of the ideas of opponents.

The work of Democritus survives thanks to being extensively cited by Aristotle and his later follower Epicurus (341 B.C.–270 B.C.). Appreciation of Epicurus' work, depends largely on the Latin epic poem "De rerum natura" (On the nature of things) (circa 94–55 B.C.) written by Lucretius in Rome. There, in book 2, he describes the shape and motion of atoms and in book 3 he states that the soul is mortal and, because of its death, it is not to be feared. This work had to await until the Italian Renaissance (1400s–1500s) to be rediscovered because it was suppressed during the Middle Ages.

26.5 Plato's Assumption Has Been Taken for Granted for Several Reasons: Atoms Have Only Been Photographed in the Last Decades

The emphasis on an unreal world suited the thinking of the Dark Ages and the static view dominating philosophy during several centuries.

For years living organisms were only divisible into organs distinguished by the naked eye. When the microscope was invented cells were discovered but initially seemed to be empty structures (Hooke 1665). It was only by 1847 that Schwann found them to contain a nucleus and a nucleolus.

The synthesis of urea in the laboratory was a revolutionary event which demonstrated that biological substances were not distinct from inorganic ones (Wöhler 1828). After 10 years Berzelius coined for the first time the term protein. Macromolecules became accepted as components of cells but had not yet been seen. Atoms were definitely out of reach. Their mass and precise size had been determined by Niels Bohr (1885–1962) but they also could not be seen even with the best light microscopes.

Only after World War II did the electron microscope become fully used in the study of the cell. As late as 1969 the first electron photographs of the macromolecule DNA were making a sensation.

Rapid technological development in the last decade enabled single atoms to be photographed with the most powerful microscopes. The electron, with a mass which is utterly small (1/1837 of a proton) is now close to being photographed (Fig. 26.1). Four centuries of research have led to the successive disposal of abstractions by converting them into realities.

The real world is becoming palpable at levels before thought to be out of reach of human capacity.

26.6 "Plato's Extreme Position Turned Out to Be a Grave Mistake Chastised by Modern Scientists"

Throughout time, Aristotle, Democritus, Epicurus, Lucretius and others had opposed Plato's interpretation of the world. Physicists have lately insisted on the necessity to dispose of his contention. Padamsee (2003) does not hesitate to put it clearly. "Plato split the cosmos into two distinct realms, the world of 'form' and the world of 'substance'". The first is the domain of pure ideals, of perfect forms such as spheres, circles, and perfect solids. The second realm is the imperfect and changing physical world composed of shadowy copies of perfect forms of the realm of ideas. "Plato's extreme position, turned out to be a grave mistake". For his arrogance "Plato has been severely chastised by modern scientists who have come to appreciate the value of empirical investigation".

26.7 The Reversal of Plato's Contention: The Real Is the Palpable World, Absolute Perfection Is a Product of Brain Imagery—There Are Two Types of Geometry

Matter, at its very onset, displays geometry and this has been kept undisturbed all the way to humans. This property is evident at three distinct levels. First, the left-handed and right-handed forms found in galaxies are visible again in the shape and function of the human body (Fig. 15.2). Second, the cells of the brain display geometric patterns

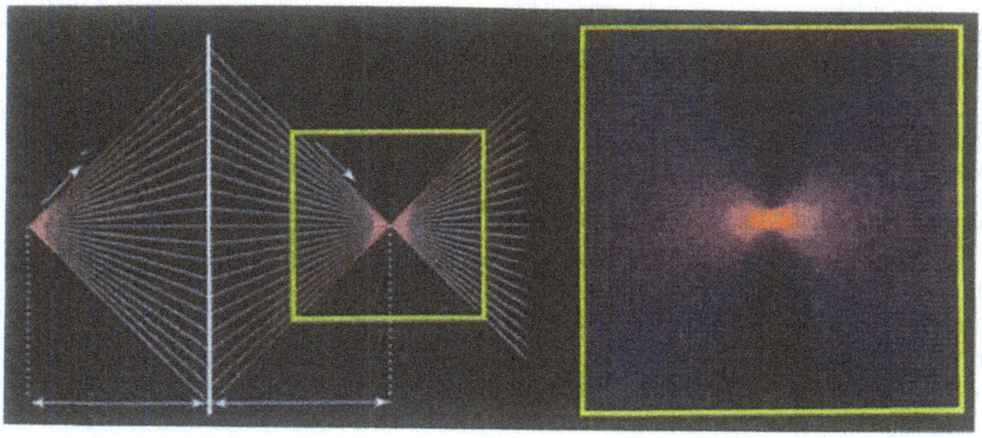

Fig. 26.1 **Focusing electrons**. An attempt has been made to focus electrons by using graphene as a lens. This material consists of single sheets of carbon atoms and these have allowed to focus electrons to extremely fine points. (*Left*) Light formerly diverging from a point source is set in reverse and converges back to a point. (*Right*) A computer simulation of electron charge density on graphene showing similar focusing

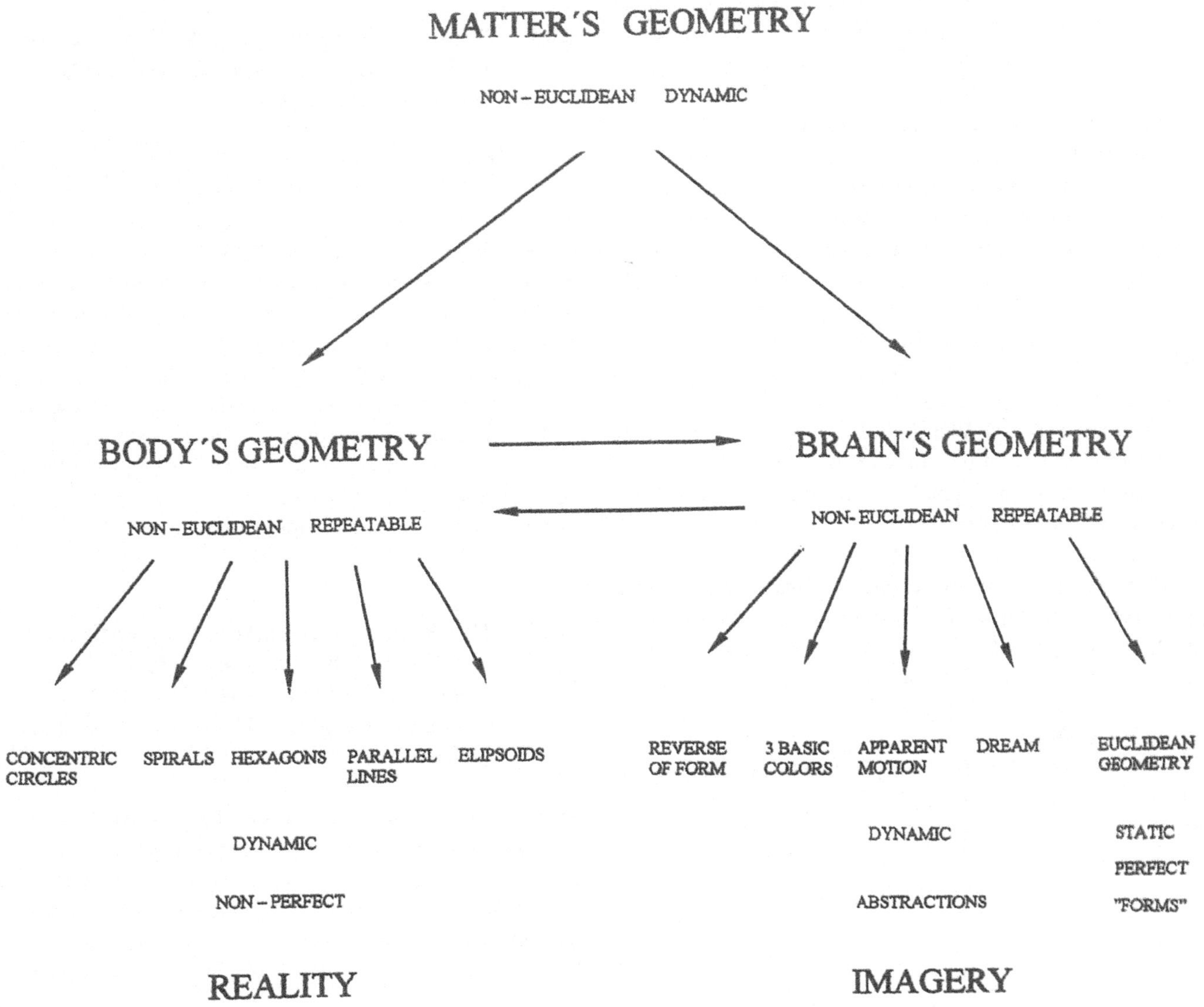

Fig. 26.2 The relationship between the geometries of matter, body and brain. Their interactions and effects are marked by *arrows*. The three *geometries* are Non-Euclidean and real since they build three-dimensional structures in which the *geometric figures* tend to form on non-flat surfaces. The atom with its many circulating electrons, occupying different orbits, has already a shape that is far from flat. The brain's and body's geometry are repeatable because they emerge in identical form when transmitted to the progeny. The geometry of matter is also repeatable, but in a simpler form, because a given atomic configuration tends to occur in a specific *crystal shape*. The geometry of the brain leads to five main types of imagery. One of these is the Euclidean geometry, which is static dealing with perfect "forms", only conceived by the brain of animals and humans. Living organisms consist of matter that exudes geometry at every level of organization. The perfect "forms" conceived by living minds are a secondary type of geometry that is imaginary. How the real and the imaginary representations are interdependent remains an open question. One of the present main preoccupations of molecular biologists is the study of the brain at the molecular level. A Pandora's box of surprising results will create a whole series of philosophical problems having unpredictable ethical consequences

that are an obligatory component of their ordered mental processes. Third, the perception of geometry is not an abstraction but has been localized in the human brain's right hemisphere (Güntürkün 2012). Following an extensive study of brain asymmetry in vertebrates, in which he studied the handedness and visual asymmetries of many species from frogs, birds, great apes to humans, this author concluded that most studies have documented the relative predominance of the right hemisphere in utilizing the geometry of the environment. Experiments in chicks and pigeons confirmed that the right hemisphere is the area of the brain where geometric processing takes place.

The evidence indicates that the geometry perceived by the brain could not exist if it was not an initial property of matter.

The finding that invertebrates, like spiders, or vertebrates, such as birds, think in geometric terms, and act building highly geometric structures, demonstrates that this process is in no way reserved to humans, but is a part of the evolution of living organisms which has been conditioned by the ordered organization of matter.

At the same time the brain discloses its imagery in several well-defined ways, by actually creating a reality of its own distinct from the actual reality.

First, there are no colors in nature. There is only electromagnetic radiation. Still most significant is that the whole spectrum of radiation is not discontinuous. There is no gap or change in the quality, of the waves that the brain chooses to conceive as colors. The wavelength of radiation follows a continuum (Fig. 24.2).

Second, movement is easily divided into real movement, occurring independently of the observer, and illusory movement that the brain builds from figures that are static.

Third, there are also two types of geometry: the real one that is an integral part of matter which is palpable in crystals (and seen also using microscopes and telescopes) and the illusory one, consisting of perfect spheres, circles and triangles, that the human brain processes using the same imagery that it employs in the invention of color and of movement (Fig. 26.2).

As Kandel et al. (1995) state "Our perceptions, as direct and precise images of the world, is an illusion".

26.8 There Is No Reason to Suppose that Evolution Finished with *Homo Sapiens*: DNA Mutations May Change Our Posture in Relation to the Universe

Humans, like any other living organism have had an evolution. The human lineage is considered to have begun with *Australopithecus afarensis* (35 million years ago). It developed into *Homo habilis*, who in turn became *Homo erectus* (1 million years ago). The fossil record contains a number of early types of *Homo sapiens* such as *Homo neanderthalensis*. There is no indication that evolution would stop at the present stage.

Our posture in face of the universe has been one of submission and at the same time one of rebellion.

On one hand we have been highly impressed by the forces of natural phenomena. Volcanos spitting fire and river currents that could not be contained, gave us a sense of our impotence in face of a mighty and uncontrollable nature. This led us to have an attitude of submission as we realized the diminutive nature of our physical and mental ability in relation to the universe.

On the other hand we rebelled, by refusing to accept this posture. We created tools of all kinds to probe the construction of the outer world and to control it.

But this type of attitude and vision of our surrounding world may change suddenly due to a mutation of a gene, or a group of genes, that may affect our mental behavior. The chemical evolution of DNA cannot be contained.

At present we can only conceive that $2 + 2 = 4$ but a mutation may liberate us from this possible imagery.

We could as well revert to a mental status, which seems to be present in most animals, in which one seems not to be preoccupied with the birth of the universe or its future. But we may mutate into an even more inquisitive species which will have an expanding mind.

In any case we are not expected to remain as *Homo sapiens* for ever.

Sources of Figures

Fig. 26.1 Pendry, J.B. 2007. Negative refraction for electrons? Science 315: 1226–1227 (Fig. on page 1226 adapted from Fig. 2 page 1253 of Cheianov, V.V. et al. 2007. The focusing of electron flow and a Veselago lens in graphene p-n junctions. Science 315: 1252–1255).

Fig. 26.2 Original. Diagram showing the relationship between Matter's geometry, Body's geometry and Brain's geometry. Not to be reproduced without complete reference to this work.

References

Abercrombie M et al (1990) The new penguin dictionary of biology. Penguin Books, London

Abrahamsen H, Stenmark H (2011) Growth signaling from inside. Science 334:611–612

Alexander T et al (2009) *Hox* genes and segmentation of the hindbrain and axial skeleton. Annu Rev Cell Dev Biol 25:431–456

Atkins PW (1987) Molecules. Scientific American Library, New York

Bachiller D et al (1994) Conservation of a functional hierarchy between mammalian and insect *Hox/Hom* genes. EMBO J 13:1930–1941

Baker R (ed) (1980) The mystery of migration. Macdonald and Jane's, London

Barnes RD (1980) Invertebrate zoology. Saunders College, Philadelphia, PA

Barth FG (2002) A spider's world: senses and behaviour. Springer, Berlin

Bassnett S (2009) On the mechanism of organelle degradation in the vertebrate lens. Exp Eye Res 88:133–139

Baylin SB, Herman JG (2000) DNA hypermethylation in tumorigenesis: epigenetics joins genetics. Trends Genet 16 (4):168–173

Beebe W (1943) Pattern and color in the cichlid fish, *Aequidens tetramerus*. Zoologica, New York Zoological Society 28 (1–23):12–16, plus plate 1

Beltrami G, Bertolucci C, Parretta A, Petrucci F, Foà A (2010) A sky polarization compass in lizards: the central role of the parietal eye. J Exp Biol 213:2048–2054

Bender W et al (1983) Molecular genetics of the bithorax complex in *Drosophila melanogaster*. Science 221:23–29

Benirschke RJ et al (1964) Chromosome studies of a donkey-Grevy zebra hybrid. Chromosoma 15:1–13

Berrill NJ, Karp G (1976) Development. McGraw-Hill, New York

Blackburn S (1996) The Oxford dictionary of philosophy. Oxford University Press, Oxford

Blaringhem L (1923) Pasteur et le transformisme. Masson et Cie. Editeurs. Libraires de l'Academie de Medecine, Paris

Bonner JT (1947) Evidence for the formation of cell aggregates by chemotaxis in the development of the slime mold *Dictyostelium discoideum*. J Exp Zool 106:1–26

Bonner JT (1959) The cellular slime molds. Princeton University Press, Princeton, NJ

Bork P, Copley R (2001) Filling in the gaps. Nature 409:818–820

Botchkarev VA et al (2012) Epigenetic regulation of gene expression in keratinocytes. J Investig Dermatol 132:2505–2521

Boyle WA, Conway CJ (2007) Why migrate? A test of the evolutionary precursor hypothesis. Am Nat 169:344–359

Brown C, Garwood MP, Williamson JE (2012) It pays to cheat: tactical deception in a cephalopod social signalling system. Biol Lett Anim Behav 2012:1–4, The Royal Society

Brunetta L, Craig CL (2010) Spider silk. Yale University Press, New Haven, CT

Brunnin N, Tsui-James EP (eds) (2003) The Blackwell companion to philosophy. Blackwell, Oxford

Burnie D (ed) (2004) Animal. Dorling Kindersley, London

Butler PIG (1999) Self-assembly of tobacco mosaic virus: the role of an intermediate aggregate in generating both specificity and speed. Philos Trans R Soc Lond B Biol Sci 354:537–550

Butler AB (2012) Evolution of the brain in vertebrates. In: Lazareva OF et al (eds) How animals see the world. Oxford University Press, Oxford, pp 419–437

Bynum WF, Browne EJ, Porter R (1983) Dictionary of the history of science. The Macmillan Press, London

Byrne O (1847) The first six books of the elements of euclid. William Pickering, London (Reprinted by Taschen, 2010 www.taschen.com)

Cakmak YO (2009) A review of the potential effect of electroacupuncture and moxibustion on cell repair and survival: the role of heat shock proteins. Acupunct Med 27(4):183–186

Callaerts P et al (1997) *Pax-6* in development and evolution. Ann Rev Neurosci 20:483–532

Campbell CBG, Hodos W (1991) The *scala naturae* revisited: evolutionary scales and anagenesis in comparative psychology. J Comp Psychol 105:211–221

Carroll RL (1988) Vertebrate paleontology and evolution. W.H. Freeman and Company, New York

Cech TR, Bass BL (1986) Biological catalysis by RNA. Annu Rev Biochem 55:599–630

Chan Y–HM, Marshall WF (2012) How cells know the size of their organelles. Science 337:1186–1189

Changeux J-P (1985) Neuronal man: the biology of mind. Oxford University Press, Oxford

Cheers G (ed) (2003) Geographica, Swedish Edition. Random House Australia, Könemann

Cheianov VV et al (2007) The focusing of electron flow and a Veselago lens in graphene *p-n* junctions. Science 315:1252–1255

Clark RW (1973) Einstein – the life and times. Hodder and Stoughton, London

Cogger HG, Zweifel RG (1998) Encyclopedia of reptiles and amphibians. Academic, San Diego, CA

Cohn MJ et al (1997) *Hox9* genes and vertebrate limb specification. Nature 387:97–101

Colbert EH (1980) Evolution of the vertebrates. Wiley, New York

Colbourne JK et al (2011) The ecoresponsive genome of *Daphnia pulex*. Science 331:555–561

Coon H, Carey G (1989) Genetic and environmental determinants of musical ability in twins. Behav Genet 19(2):183–193

Counis MF et al (1998) Analysis of nuclear degradation during lens cell differentiation. Cell Death Differ 5:251–261

Crease RP (2009) A brief guide to the great equations. Robinson, Constable and Robinson, London

Creuzet S et al (2005) Patterning the neural crest derivatives during development of the vertebrate head: insights from avian studies. J Anat 207:447–459

Cuénot L (1932) La Genese des Especes Animales. Libairie Felix Alcan, Paris

A. Lima-de-Faria, *Molecular Origins of Brain and Body Geometry*,
DOI 10.1007/978-3-319-06056-9, © Springer International Publishing Switzerland 2014

Daeschler T, Shubin N (2011) Fossil discovery and the origins of tetrapods. In: Losos JB (ed) In the light of evolution. Roberts and Company, Greenwood Village, CO, pp 297–310
Damasceno PF et al (2012) Predictive self-assembly of polyhedra into complex structures. Science 337:453–457
Damasio A (1994) Descartes' error. Putnam, New York
Damasio A (1999) The feeling of what happens. Harcourt Brace and Company, New York
de la Roche MA et al (2002) Signaling pathways regulating *Dictyostelium* myosin II. J Muscle Res Cell Motil 23:703–718
del Hoyo J et al (eds) (1992–2011) Handbook of the birds of the world, vol 1–16. Lynx Edicions, Barcelona
Delius C et al (2000) Filosofins Historia från Antiken till Idag. Könemann, Köln
Drayna D et al (2001) Genetic correlates of musical pitch recognition in humans. Science 291:1969–1972
Dreller C, Kirchnert WH (1993) How honeybees perceive the information of the dance language. Naturwissenschaften 80:319–321
Dubertret L et al (1987) Les peaux artificielles vivantes. La Recherche 18:149–155
Dudits D et al (1976) Fusion of human cells with carrot protoplasts induced by polyethylene glycol. Hereditas 82:121–124
Duester G (2008) Retinoic acid synthesis and signaling during early organogenesis. Cell 134:921–931
Eccles JC (1989) Evolution of the brain: creation of the self. Routledge, London
Eckert R, Randall D (1978) Animal physiology. W.H. Freeman and Company, San Francisco, CA
Egri A et al (2012) Polarotactic tabanids find striped patterns with brightness and/or polarization modulation least attractive: an advantage of zebra stripes. J Exp Zool 215:736–745
Elphick J (ed) (2007) The Atlas of bird migration. Natural History Museum, London
Ericson PGP, Sjögren H (2004) Boken om Göken. Atlantis, Stockholm
Erlande-Brandenburg A (1995) The cathedral builders of the middle ages. Thames and Hudson, London
Etchberger CR et al (1993) Environmental and maternal influences on embryonic pigmentation in a turtle (*Trachemys scripta elegans*). J Zool (Lond) 230:529–539
Feil R, Khosla S (1999) Genomic imprinting in mammals an interplay between chromatin and DNA methylation? Trends Genet 15 (11):431–434
Feringa BL (2002) Cited by Jones N (2002) Scientists learn to tell left from right. New Scientist (Feb 9):19
Finlay BL, Darlington RB (1995) Linked regularities in the development and evolution of mammalian brains. Science 268:1578–1584
Finlay BL et al (2001) Developmental structure in brain evolution. Behav Brain Sci 24:263–308
Finn JK et al (2009) Defensive tool use in a coconut-carrying octopus. Curr Biol 19(23):R1069–R1070
Flutre F (1935) Pascal. Pensées et Opuscules Philosophiques. Classiques Illustrés Vaubourdolle, Librairie Hachette, Paris
Foelix RF (1982) Biology of spiders. Harvard University Press, Cambridge, MA
Foelix RF (1996) Biology of spiders. Oxford University Press, Oxford
Foelix RF (2011) The biology of spiders. Oxford University Press, Oxford
Folkman J, Haudenschild C (1980) Angiogenesis in vitro. Nature 288:551–556
Fraenkel-Conrat H (1962) Design and function at the threshold of life: the viruses. Academic, New York
Freeman S, Herrow JC (2004) Evolutionary analysis. Pearson, Prentice Hall, NJ
Friday A, Ingram DS (1985) The Cambridge encyclopedia of life sciences. Cambridge University Press, Cambridge
Futahashi R et al (2012) Redox alters yellow dragonflies into red. Proc Natl Acad Sci USA 109:12626–12631
Gahr M et al (1993) Estrogen receptors in the avian brain: survey reveals general distribution and forebrain areas unique to songbirds. J Comp Neurol 327:112–122
Galtsoff PS (1923) The amoeboid movement of dissociated sponge cells. Biol Bull 45:153–161
Garb JE et al (2006) Silk genes support the single origin of orb webs. Science 312:1762
Gatesy J et al (2001) Extreme diversity, conservation, and convergence of spider silk fibroin sequences. Science 291:2603–2605
Gehring WJ (1998) Master control genes in development and evolution: the homeobox story. Yale University Press, New Haven, CT
Gehring WJ (2002) The genetic control of eye development and its implications for the evolution of the various eye-types. Int J Dev Biol 46:65–73
Gerisch G (1968) Cell aggregation and differentiation in Dictyostelium. In: Moscona AA, Monroy A (eds) Current topics in developmental biology. Academic, New York
Gilbert SF (2000) Developmental biology. Sinauer Associates, Sunderland, MA
Gill FB (2003) Ornithology. W.H. Freeman and Company, New York
Gillin JC et al (1985) Cholinergic mechanisms in REM sleep. In: Vauquier A et al (eds) Sleep: neurotransmitters and neuromodulators. Raven, New York, pp 29–42
Glashow S (1997) Blessed is the weak. New Scientist 1997:28–31
Golan DE et al (eds) (2005) Principles of pharmacology. Lippincott Williams and Wilkins, Philadelphia, PA
Grassé PP (1977) Larousse animal portraits. Hamlyn, London
Gray AP (1972) Mammalian hybrids. Commonwealth Agricultural Bureaux, Slough
Greewood AD et al (1999) Nuclear DNA sequences from late *Pleistocene megafauna*. Mol Biol Evol 16(11):1466–1473
Gregg JH, Yu NY (1975) *Dictyostelium* aggregate-less mutant plasma membranes. Exp Cell Res 96:283–286
Grewal SIS, Moazed D (2003) Heterochromatin and epigenetic control of gene expression. Science 301:798–802
Grout DJ, Palisca CV (2001) A history of western music. W.W. Norton and Company, New York
Grzimek B (2004) Hutchins M et al (eds) Grzimek's animal life encyclopedia. Gale Group, Farmington Hills, MI
Güntürkün O (2012) Brain asymmetry in vertebrates. In: Lazareva OF (ed) How animals see the world. Oxford University Press, Oxford, pp 501–514
Haesler S et al (2007) Incomplete and inaccurate vocal imitation after knockdown of *FoxP2* in songbird basal ganglia nucleus area X. PLoS Biol 5:2885–2897
Hager R et al (2012) Genetic architecture supports mosaic brain evolution and independent brain-body size regulation. Nat Commun 3, 1079, 1–5
Halder GP et al (1995) Induction of ectopic eyes by targeted expression of the *eyeless* gene in *Drosophila*. Science 267:1788–1792
Halliday T, Adler K (eds) (2004) The new encyclopedia of reptiles and amphibians. Oxford University Press, Oxford
Halstead LB (1978) The evolution of the mammals. Peter Lowe, Eurobook Limited, Lambert Beverly
Hargittai I, Hargittai M (1995) Symmetry through the eyes of a chemist. Plenum, New York
Harris N (2003) The history of ancient Egypt. Chancellor, London
Harrison C (1975) A field guide to the nests, eggs and nestlings of British and European birds. Collins, London

Hattstein M (2011) Svunna Riken (Swedish Translation). Parragon, New York
Helgen KM (2011) The mammal family tree. Science 334:458–459
Herrera AM et al (2013) Developmental basis of phallus reduction during bird evolution. Curr Biol 23:1–10, June 17, 2013
Hetcht S (1942) The quantum relations of vision. J Opt Soc Am 32:42–49
Hey T, Walters P (2003) The new quantum universe. Cambridge University Press, Cambridge
Hill RJ, Stollar BD (1983) Dependence of Z DNA antibody binding to polytene chromosomes on acid fixation and DNA torsional strain. Nature 305:338–340
Hill RW et al (2004) Animal physiology. Sinauer Associates, Sunderland, MA
Hobson JA (2009) *REM* sleep and dreaming towards a theory of protoconsciousness. Nat Rev Neurosci 10:803–813
Hochberg J (1968) In the mind's eye. In: Haber RN (ed) Contemporary theory and research in visual perception. Holt, Rinehart and Winston, New York, pp 309–331
Hoekstra HE (2011) From Darwin to DNA: the genetic basis of color adaptations. In: Losos JB (ed) In the light of evolution. Roberts and Company, Greenwood Village, CO, pp 277–295
Hoekstra HE et al (2006) A single amino acid mutation contributes to adaptative beach mouse color pattern. Science 313:101–104
Holley SA (2007) The genetics and embryology of zebrafish metamerism. Dev Dyn 236:1422–1449
Holyoak DT (2001) Nightjars and their allies. Oxford University Press, Oxford
Honour H, Fleming J (2002) A world history of art. Laurence King, London
Hornblower S, Spawforth A (1999) The Oxford classical dictionary. Oxford University Press, Oxford
Hsu TC, Benirschke K (1971) In: Hsu TC, Benirschke K (eds) An atlas of mammalian chromosomes, vol 6. Folio 291. Springer, Heidelberg
Huang W-C et al (2012) Synthesis of Cu_2O nanocrystals from cubic to rhombic dodecahedral structures and their comparative photocatalytic activity. J Am Chem Soc 134:1261–1267
Hubel DH, Wiesel TN (1959) Receptive fields of single neurons in the cat's striate cortex. J Physiol 148:574–591
Hubel DH, Wiesel TN (1962) Receptive fields, binocular interaction and functional architecture in the cat's visual cortex. J Physiol 160:106–154
Hubel DH, Wiesel TN (1979) Brain mechanisms of vision. Sci Am 241 (3):150–162
Hunt DM et al (2009) Evolution and spectral tuning of visual pigments in birds and mammals. Philos Trans R Soc B 364:2941–2955
Ian D et al (2002) The Cambridge dictionary of scientists. Cambridge University Press, Cambridge
Isaac A et al (1998) Tbx genes and limb identity in chick embryo development. Development 125:1867–1875
Jex AR et al (2011) *Ascaris suum* draft genome. Nature 479:529–533
Ji P et al (2011) Formation of mammalian erythrocytes: chromatin condensation and enucleation. Trends Cell Biol 21:409–415
Kandel ER et al (eds) (1995) Essentials of neural science and behavior. Prentice Hall International, London
Kardong KV (1995) Vertebrates. Wm. C. Brown, Dubuque, IW
Karmanova IG (1982) Evolution of sleep (Translation from Russian). Karger, Basel
Kegel WK, van der Schoot P (2006) Physical regulation of the self-assembly of tobacco mosaic virus coat protein. Biophys J 91:1501–1512
Kelsh RN et al (2009) Stripes and belly-spots – a review of pigment cell morphogenesis in vertebrates. Semin Cell Dev Biol 20:90–104
Kenny A (2001) The Oxford illustrated history of western philosophy. Oxford University Press, Oxford
Kim EB et al (2011) Genome sequencing reveals insights into physiology and longevity of the naked mole rat. Nature 479:223–227
King BL et al (2011) A natural deletion of the *HoxC* cluster in elasmobranch fishes. Science 334:1517
Kondo T et al (2010) Small peptides switch the transcriptional activity of shavenbaby during *Drosophila* embryogenesis. Science 329:336–339
Koretz JF, Handelman GH (1988) How the human-eye focuses. Sci Am 259:92–99
Koyanagi M et al (2008) Jellyfish vision starts with *cAMP* signaling mediated by *opsin-*G_s cascade. Proc Natl Acad Sci USA 105:15576–15580
Kraus P, Lufkin T (2006) *Dlx* homeobox gene control of mammalian limb and craniofacial development. Am J Med Genet Part A 140A:1366–1374
Kubikova L et al (2010) Dopamine receptors in a songbird brain. J Comp Neurol 518:741–769
Kuczmarski ER, Spudich JA (1980) Phosphorylation of myosin heavy chain *Dictyostelium* regulates self-assembly. In: 20th annual meeting American social cell biology. J Cell Biol 87:227A
Kudo RR (1971) Protozoology. Charles C. Thomas, Springfield, IL
Kuehn MR et al (1987) A potential animal model for Lesch-Nyhan syndrome through introduction of *HPRT* mutations into mice. Nature 326:295–298
LaBerge S, Rheingold H (1990) Exploring the World of lucid dreaming. Ballantine, New York
Labra A et al (2010) Evolution of the third eye: a phylogenetic comparative study of parietal-eye size as an ecophysiological adaptation in *Liolaemus* lizards. Biol J Linnean Soc 101:870–883
Lackie JM, Dow JAT (eds) (1995) The dictionary of cell biology. Academic; Harcourt Brace and Company, London
Lagercrantz H, Olson L (2007) Fostrets och barnets hjärna. In: Olson L (ed) Hjärnan. Karolinska Institutet University Press, Stockholm, pp 147–166
Lamarck, J. 1809. *Philosophie Zoologique* ou Exposition des Considérations Relatives a l'Histoire Naturelle des Animaux. Librairie Schleicher Frères, Paris (1868)
Langille MR et al (2012) Stepwise evolution of spherical seeds into 20-fold twinned icosahedra. Science 337:954–957
Lawrence PA (1992) The making of a fly: the genetics of animal design. Blackwell, Oxford
Lazareva OF et al (eds) (2012) How animals see the world. Oxford University Press, Oxford
Le Douarin NM et al (2012) The neural crest is a powerful regulator of pre-otic brain development. Dev Biol 366:74–82
Lehner N, Howk JC (2011) A reservoir of ionized gas in the galactic halo to sustain star formation in the Milky Way. Science 334:955–958
Lehninger AL (1975) Biochemistry. Worth Publication, New York
Leick V et al (1979) Scanning electron microscopy of the nuclei and nucleoli in Tetrahymena. Hereditas 90:219–226
Leonard JA et al (2005) A rapid loss of stripes: the evolutionary history of the extinct quagga. Biol Lett 1:291–295
Li H-B, Henning T (2011) The alignment of molecular cloud magnetic fields with the spiral arms in *M33*. Nature 479:499–501
Libbrecht K (2011) The snowflake designer. Nature 480:453–454
Liljas A (2004) Structural aspects of protein synthesis. World Scientific, Singapore
Lima-de-Faria A (1973) Equations defining the position of ribosomal cistrons in the eukaryotic chromosome. Nature 241:136–139
Lima-de-Faria A (1977) Genetic implications of the fusion of human cells with plant protoplasts. In: de la Chapelle A, Sorsa M (eds) Chromosomes today, vol 6. Elsevier, Amsterdam, pp 273–279
Lima-de-Faria A (1983) Molecular evolution and organization of the chromosome. Elsevier, Amsterdam

Lima-de-Faria A (1988) Evolution without selection: form and function by autoevolution. Elsevier, Amsterdam

Lima-de-Faria A (1995) Biological periodicity: its molecular mechanism and evolutionary implications. JAI Press, Greenwich, CT

Lima-de-Faria A (2003) One hundred years of chromosome research and what remains to be learned. Kluwer, Dordrecht

Lima-de-Faria A (2008) Praise of chromosome "Folly": confessions of an untamed molecular structure. World Scientific, Singapore

Lima-de-Faria A (2012) Molecular geometry of body pattern in birds. Springer, Berlin

Lima-de-Faria A et al (1977) Fusion of human cells with *Haplopappus* protoplasts by means of Sendai virus. Hereditas 87:57–66

Lima-de-Faria A et al (1983) Fusion of bull and human sperm with *Nicotiana* protoplasts. Hereditas 99:191–201

Lincoln D (2012) The inner life of quarks. Sci Am 2012:36–43

Line L et al (eds) (1983) The Audubon society book of insects. A Chanticleer Press Edition. Harry N. Abrams, New York

Liu Y, Xiao A (2011) Epigenetic regulation in neural crest development. Birth Defects Res (Part A) 91:788–796

Livio M (2012) Why symmetry matters. Nature 490:472–473

Lumsden A, Keynes R (1989) Segmental patterns of neuronal development in the chick hindbrain. Nature 337:424–428

Lycan WG (2003) Philosophy of mind. In: Bunnin N, Tsui-James EP (eds) The Blackwell companion to philosophy. Blackwell, Oxford, pp 173–196

Macdonald D (ed) (2002) The new encyclopedia of mammals. Oxford University Press, Oxford

Macfarlane RJ et al (2011) Nanoparticle superlattice engineering with DNA. Science 334:204–208

Mackay AL (1987) Quasicrystals and amorphous materials. J Non-Cryst Solids 97–98:55–62

Marshall H et al (1996) Retinoids and *Hox* genes. FASEB J 10 (9):969–978

Martin G et al (1989) Epidermal focusing and the light microenvironment within leaves of *Medicago sativa*. Physiol Plant 76:485–492

Masterton WL, Hurley CN (1993) Chemistry: principles and reactions. Saunders College, Philadelphia, PA

Matzke M, Matzke AJM (2003) RNAi extends its reach. Science 301:1060–1061

Mazurs EG (1974) Graphic representation of the periodic system during one hundred years. University of Alabama, Birmingham, AL

McGinnis W, Kuziora M (1994) The molecular architects of body design. Sci Am 270:36–42

McGrath KP, Butler MM (1997) Self-assembling protein systems: a model for materials science. In: McGrath K, Kaplan D (eds) Protein-based materials, Bioengineering materials. Birkhäuser, Boston, pp 251–279

Merchant SS et al (2007) The *Chlamydomonas* genome reveals the evolution of key animal and plant functions. Science 318:245–251

Meredith RW et al (2011) Impacts of the cretaceous terrestrial revolution and K Pg extinction on mammal diversification. Science 334:521–524

Mintz B (1978) Mutagenized teratocarcinoma cells as probes of mammalian differentiation. In: Plenary sessions symposia, abstracts. XIV international congress of genetics, Moscow, p 65

Mintz B, Illmensee K (1975) Normal genetically mosaic mice produced from malignant teratocarcinoma cells. Proc Natl Acad Sci USA 72:3585–3589

Mittermeier RA et al (eds) (2013) Handbook of the mammals of the world, vol 3. Lynx Edicions, Barcelona

Montgomery SH et al (2010) Reconstructing the ups and downs of primate brain evolution: implications for adaptive hypotheses and *Homo floresiensis*. BMC Biol 8(9):1–19

Moore P (1978) The Rand McNally new concise atlas of the universe. Mitchell Beazley, Chicago, IL

Morgan TH (1934) Embryology and genetics. Columbia University Press, New York

Müller WEG et al (1976) Species-specific aggregation factor in sponges IV. Exp Cell Res 98:31–40

Müntzing A (1961) Genetic research: a survey of methods and main results. LTs Förlag, Stockholm

Narita Y, Rijli FM (2009) *Hox* genes in neural patterning and circuit formation in the mouse hindbrain. Curr Top Dev Biol 88:139–167

Neumeyer C (2012) Color vision in goldfish and other vertebrates. In: Lazareva OF et al (eds) How animals see the world. Oxford University Press, Oxford

Nir Y, Tononi G (2009) Dreaming and the brain: from phenomenology to neurophysiology. Trends Cogn Sci 14:88–100

Nomura M (1973) Assembly of bacterial ribosomes. Science 179:864–873

Norman MD et al (2001) Dynamic mimicry in an Indo-Malayan octopus. Proc R Soc Lond B 268:1755–1758

Norris DP, Grimes DT (2012) Cilia discern left from right. Science 338:206–207

North J (2008) Cosmos: an illustrated history of astronomy and cosmology. University Chicago Press, Chicago, IL

O'Connor DJ (ed) (1985) A critical history of western philosophy. Macmillan, London

Ohno S (1972) So much "Junk" DNA in our genome. In: Smith HH (ed) Evolution of genetic systems. Gordon and Breach, New York, pp 366–370

Ohno S et al (1978) Testicular cells lysostripped of H-Y antigen organize ovarian follicle-like aggregates. Cytogenet Cell Genet 20:351–364

Orr R (1986) Mammals of Britain and Europe. Peerage Books, London

Padamsee HS (2003) Unifying the universe. Institute of Physics, Bristol

Pagels HR (1982) The cosmic code: quantum physics as the language of nature. Michael Joseph, London

Parker G (ed) (1986) The times: the world. An illustrated history. Times Books, London

Pauling L (1949) General chemistry. W.H. Freeman and Company, San Francisco, CA

Paxton JR, Eschmeyer WN (eds) (1998) Encyclopedia of fishes (natural world). Academic, San Diego, CA

Payne RB (2005) The cuckoos. Oxford University Press, Oxford

Pearce JM (2008) Animal learning and cognition, an introduction. Psychology Press; Taylor and Francis Group, New York

Pennisi E (2003) Gene counters struggle to get the right answer. Science 301:1040–1041

Petie R (2012) Visually guided swimming in box jellyfish. Doctor's thesis, Department of Biology, Lund University, Lund, Sweden

Pitt VH (ed) (1988) The penguin dictionary of physics. Penguin Books, London

Plutarch (1965) Makers of Rome. Penguin Classics, London

Polilov AA (2012) The smallest insects evolve anucleate neurons. Arthropod Struct Dev 41:29–34

Popper KR (1976) Darwinism is a metaphysical research programme. In: Unended quest: an intellectual autobiography. Fontana, London

Preston-Mafham R (1996) Den Stora Boken om Spindlar. Streiffert Förlag AB, Stockholm

Preuss TM (2001) The discovery of cerebral diversity: an unwelcome scientific revolution. In: Falk D, Gibson K (eds) Evolutionary anatomy of the primate cerebral cortex. Cambridge University Press, Cambridge

Pugsley AP (1989) Protein targeting. Academic, New York

Puller C, Haverkamp S (2011) Bipolar cell pathways for color vision in non-primate dichromats. Vis Neurosci 28:51–60

Pulli K et al (2008) Genome-wide linkage scan for loci of musical aptitude in Finnish families: evidence for a major locus at 4q22. J Mol Genet 45:451–456

Purves D et al (eds) (2012) Neuroscience. Sinauer Associates, Sunderland, MA

Putnam NH et al (2008) The amphioxus genome and the evolution of the chordate karyotype. Nature 453:1064–1072

Quiring R et al (1994) Homology of the *eyeless* gene of *Drosophila* to the *Small eye* gene in mice and *Aniridia* in humans. Science 265:785–789

Ramachandran VS et al (1996) Rapid adaptive camouflage in tropical flounders. Nature 379:815–818

Ramón y Cajal S (1911) Histologie du Systeme Nerveux de l'Homme et des Vertebres. Maloine, Paris

Ridgway SH (1986) Physiological observations on dolphin brains. In: Schusterman RJ, Thomas JA, Wood FG (eds) Dolphin cognition and behavior: a comparative approach. Lawrence Erlbaum Association, Hillsdale, pp 31–59

Rilling JK (2001) Allometric departures for the human brain provide insights into hominid brain evolution. Behav Brain Sci 24 (2):292–293

Rizzoti K, Lovell-Badge R (2011) Organ recital in a dish. Nature 480:44–45

Robertson B (2009) A celebration of the 50th anniversary of David Hubel and Torsten Wiesel's *Receptive fields of single neurones in the cat's striate cortex*. J Physiol 587(12):2721–2732

Rodriguez-Esteban C et al (1999) The T-box genes *Tbx4* and *Tbx5* regulate limb outgrowth and identity. Nature 398:814–818

Röhl R, Nierhaus KN (1982) Assembly map of the large subunit (50S) of *Escherichia coli* ribosomes. Proc Natl Acad Sci USA 79:729–733

Romer J (2007) The great pyramid. Cambridge University Press, Cambridge

Romer AS, Parsons TS (1978) The vertebrate body. W.B. Saunders, Philadelphia, PA

Rowe JW et al (2006) Effect of substrate color on pigmentation in midland painted turtles (*Chrysemys picta marginata*) and red-eared slider turtles (*Trachemys scripta elegans*). J Herpetol 40:358–364

Saenger W (1988) Principles of nucleic acid structure. Springer, New York

Salter L (1978) The gramophone guide to classical composers and recordings. Salamander Books, London

Samant RS, Workman P (2012) Choose your protein partners. Nature 490:351–352

Scally A et al (2012) Insights into hominid evolution from the gorilla genome sequence. Nature 483:169–174

Scarre C (ed) (1991) Past worlds: the times Atlas of archaeology. Times Books, London

Scerri ER (2007) The periodic table. Oxford University Press, Oxford

Scharf C (2012) Gravity's engines. Scientific American; Farrar, Straus and Giroux, New York

Schartau JM (2010) Morphological and physiological adaptations in cichlid crystalline lenses. Doctor's thesis, Department of Biology, Lund University, Lund

Schmalhausen II (1949) Factors of evolution: the theory of stabilizing selection. The Blakiston Company, Philadelphia, PA

Schoenemann PT (2001) Brain scaling, behavioral ability and human evolution. Behav Brain Sci 24(2):293–295

Scholes PA (1977) The concise Oxford dictionary of music. Oxford University Press, Oxford

Schwab IR, O'Connor GR (2005) The lonely eye. Br J Ophtalmol 89:256

Sears RL (1981) Stellar energy sources and evolution. In: Lerner RG, Trigg GL (eds) Encyclopedia of physics. Addison-Wesley, Reading, MA, pp 960–961

Seifert AW et al (2012) Skin shedding and tissue regeneration in African spiny mice (*Acomys*). Nature 489:561–566

Sharp DWA (1988) The Penguin dictionary of chemistry. Penguin Books, London

Shechtman D et al (1984) Metallic phase with long-range orientational order and no translational symmetry. Phys Rev Lett 53:1951–1953

Shen P et al (1995) An atlas of aromatase mRNA expression in the zebra finch brain. J Comp Neurol 360:172–184

Shettleworth SJ (1998) Cognition, evolution and behaviour. Oxford University Press, Oxford

Solms M (2000) Dreaming and *REM* sleep are controlled by different brain mechanisms. Behav Brain Sci 23:843–850

Springer MS et al (2004) Molecules consolidate the placental mammal tree. Trends Ecol Evol 19:430–437

Stafford RE (1980) Hereditary components of musical aptitude: a review. Behav Genet 10(5):498–499

Steinhardt PJ (1990) Quasicrystals: a new form of matter. Endeavour New Ser 14:112–116

Strachan T, Read AP (2000) Human molecular genetics. Wiley-Liss, New York

Striedter GF (2004) Brain evolution. In: Paxinos G, Mai JK (eds) The human nervous system. Elsevier, New York, pp 3–21

Stryer L (1981) Biochemistry. W.H. Freeman and Company, New York

Sturtevant MH (1923) Inheritance of direction of coiling in *Limnaea*. Science 58:269–270

Suga H et al (2011) Self-formation of functional adeno-hypophysis in three-dimensional culture. Nature 480:57–63

Tachibana M et al (2013) Human embryonic stem cells derived by somatic cell nuclear transfer. Cell 153:1–11

Taipale M et al (2012) Quantitative analysis of Hsp90-client interactions reveals principles of substrate recognition. Cell 150:987–1001

Terakita A (2005) The opsins. Genome Biol 6:213 (1–9)

Thompson, D'Arcy W (1917) On growth and form, vols 1 and 2. Cambridge University Press, Cambridge (Reprinted 1952)

Travesset A (2011) Self-assembly enters the design era. Science 334:183–184

Tri M et al (2010) Intrinsically photosensitive retinal ganglion cells. Physiol Rev 90:1547–1581

Ullrich-Lüter EM et al (2011) Unique system of photoreceptors in sea urchin tube feet. Proc Natl Acad Sci USA 108:8367–8372

van Delden RA, Feringa BL (2002) Colour indicator for enantiomeric excess and assignment of the configuration of the major enantiomer of an amino acid ester. J R Soc Chem Chem Commun 2002:174–175

Varshney RK et al (2012) Draft genome sequence of pigeon pea (*Cajanus cajan*) an orphan legume crop of resource-poor farmers. Nat Biotechnol 30:83–91

Veil C (1938) L'influence du milieu sur la couleur des animaux. In: Bonnardel R et al (eds) La Biologie. Palais de la Decouverte, Masson et Cie, Paris, pp 49–56

Vester M (2004) En Värld av Nyanser. Om Synestesi. Studentlitteratur, Lund

Vogelmann TC et al (1989) Photosynthetic light gradients and spectral regime within leaves of *Medicago sativa*. Philos Trans R Soc Lond B 323:411–421

Volonteri M (2012) The formation and evolution of massive black holes. Science 337:544–547

von Baeyer HC (1986) Physics finds a childhood. The Sciences 8–10

von Denffer D et al (1971) Strasburger's textbook of botany. Longman, London

Wada S et al (2012) Expression of UV-sensitive parapinopsin in the iguana parietal eyes and its implication in UV-sensitivity in vertebrate pineal-related organs. PLoS ONE 7(6):e39003, p 7

Wakayama S et al (2008) Production of healthy cloned mice from bodies frozen at −20 °C for 16 years. Proc Natl Acad Sci USA 105(45):17318–17322

Wang D, Fernandez-Martinez A (2012) Order from disorder. Science 337:812–813

Wang L et al (2012) Long-range ordered carbon clusters: a crystalline material with amorphous building blocks. Science 337:825–828
Wang Y et al (2012) Colloids with valence and specific directional bonding. Nature 491:51–55
Warren WC et al (2008) Genome analysis of the platypus reveals unique signatures of evolution. Nature 453:175–183
Wellik DM (2009) *Hox* genes and vertebrate axial pattern. Curr Top Dev Biol 88:257–277
Wells M (1983) Cephalopods do it differently. New Scientist 100:332–337
West MJ, King AP (1990) Mozart's starling. Am Sci 78:106–114
Wilmut I et al (1997) Viable offspring derived from fetal and adult mammalian cells. Nature 385:810–813
Wilson DE, Mittermeier RA (eds) (2009) Handbook of the mammals of the world. vol 1. Lynx Edicions, Barcelona
Wilson DE, Mittermeier RA (eds) (2011) Handbook of the mammals of the world, vol 2. Lynx Edicions, Barcelona
Wirth E, Barth FG (1992) Forces in the spider orb web. J Comp Physiol A 171:359–371
Withers PC (1992) Comparative animal physiology. Saunders College, New York
Wride MA (2000) Minireview: apoptosis as seen through a lens. Apoptosis 5:203–209
Wurtman RJ, Axelrod J (1965) The pineal gland. Sci Am 213(1):50–60
Wuster DH, Atkin NB (1972) Muntjac chromosomes: a new karyotype for *Muntiacus muntjak*. Experiencia 28:972–973
Xiao C et al (2012) Dodecagonal tiling in mesoporous silica. Nature 487:349–353
Yang P (2012) Crystal cuts on the nanoscale. Nature 482:41–42
Yarbus AL (1967) Eye movements and vision (Translated by Haig B). Plenum, New York
Yong E (2011) Master of illusion. Nature 480:168–170
Yoshiba S et al (2012) Cilia at the node of mouse embryos sense fluid flow for left-right determination via *Pkd2*. Science 338:226–231
Young T, Deschamps J (2009) *Hox*, *Cdx*, and anteroposterior patterning in the mouse embryo. Curr Top Dev Biol 88:235–255
Young ND et al (2011) The *Medicago* genome provides insight into the evolution of rhizobial symbioses. Nature 480:520–523
Zenzes MT et al (1978) Studies on the function of H-Y antigen: dissociation and reorganization experiments on rat gonadal tissue. Cytogenet Cell Genet 20:365–372
Zhang U et al (1993) Ectopic *HoxA-1* induces rhombomere transformation in mouse hindbrain. Development 120:2431–2442
Zoncu R et al (2011) *mTORC1* senses lysosomal amino acids through an inside-out mechanism that requires the vascuolar H^{+}*-ATPase*. Science 334:678–683

The manufacturer's authorised representative in the EU is Springer Nature Customer Service Centre GmbH, Europaplatz 3, 69115 Heidelberg, Germany. If you have any concerns regarding our products, please contact ProductSafety@springernature.com

Printed and bound by CPI Group (UK) Ltd, Croydon, CR0 4YY

30/09/2026

02225612-0001